BURLEIGH DODDS SCIENCE: INSTANT INSIGHTS

NUMBER 97

Optimising quality attributes in horticultural products

Published by Burleigh Dodds Science Publishing Limited
82 High Street, Sawston, Cambridge CB22 3HJ, UK
www.bdspublishing.com

Burleigh Dodds Science Publishing, 1518 Walnut Street, Suite 900, Philadelphia, PA 19102-3406, USA

First published 2024 by Burleigh Dodds Science Publishing Limited
© Burleigh Dodds Science Publishing, 2024. All rights reserved.

This book contains information obtained from authentic and highly regarded sources. Reprinted material is quoted with permission and sources are indicated. Reasonable efforts have been made to publish reliable data and information but the authors and the publisher cannot assume responsibility for the validity of all materials. Neither the authors nor the publisher, nor anyone else associated with this publication shall be liable for any loss, damage or liability directly or indirectly caused or alleged to be caused by this book.

No part of this publication may be reproduced, stored in a retrieval system or transmitted in any form or by any means electronic, mechanical, photocopying, recording or otherwise without the prior written permission of the publisher.

The consent of Burleigh Dodds Science Publishing Limited does not extend to copying for general distribution, for promotion, for creating new works, or for resale. Specific permission must be obtained in writing from Burleigh Dodds Science Publishing Limited for such copying.

Permissions may be sought directly from Burleigh Dodds Science Publishing at the above address. Alternatively, please email: info@bdspublishing.com or telephone (+44) (0) 1223 839365.

Trademark notice: Product or corporate names may be trademarks or registered trademarks and are used only for identification and explanation, without intent to infringe.

Notice
No responsibility is assumed by the publisher for any injury and/or damage to persons or property as a matter of product liability, negligence or otherwise, or from any use or operation of any methods, products, instructions or ideas contained in the material herein.

British Library Cataloguing in Publication Data
A catalogue record for this book is available from the British Library

ISBN 978-1-80146-667-7 (Print)
ISBN 978-1-80146-668-4 (ePub)

DOI: 10.19103/9781801466684

Typeset by Deanta Global Publishing Services, Dublin, Ireland

Contents

© Burleigh Dodds Science Publishing Limited, 2024. All rights reserved.

© Burleigh Dodds Science Publishing Limited, 2024. All rights reserved.

Series list

Title	Series number
Sweetpotato	01
Fusarium in cereals	02
Vertical farming in horticulture	03
Nutraceuticals in fruit and vegetables	04
Climate change, insect pests and invasive species	05
Metabolic disorders in dairy cattle	06
Mastitis in dairy cattle	07
Heat stress in dairy cattle	08
African swine fever	09
Pesticide residues in agriculture	10
Fruit losses and waste	11
Improving crop nutrient use efficiency	12
Antibiotics in poultry production	13
Bone health in poultry	14
Feather-pecking in poultry	15
Environmental impact of livestock production	16
Sensor technologies in livestock monitoring	17
Improving piglet welfare	18
Crop biofortification	19
Crop rotations	20
Cover crops	21
Plant growth-promoting rhizobacteria	22
Arbuscular mycorrhizal fungi	23
Nematode pests in agriculture	24
Drought-resistant crops	25
Advances in detecting and forecasting crop pests and diseases	26
Mycotoxin detection and control	27
Mite pests in agriculture	28
Supporting cereal production in sub-Saharan Africa	29
Lameness in dairy cattle	30
Infertility and other reproductive disorders in dairy cattle	31
Alternatives to antibiotics in pig production	32
Integrated crop-livestock systems	33
Genetic modification of crops	34

© Burleigh Dodds Science Publishing Limited, 2024. All rights reserved.

© Burleigh Dodds Science Publishing Limited, 2024. All rights reserved.

© Burleigh Dodds Science Publishing Limited, 2024. All rights reserved.

© Burleigh Dodds Science Publishing Limited, 2024. All rights reserved.

Chapter 1

Developing tomato varieties with improved flavour

M. Causse, E. Albert and C. Sauvage, INRA, France

1 Introduction

Today, tomato flavour is a key issue for tomato breeders. Over the last century, tomato breeders have improved tomato yield and yield stability and have adapted to diverse growth conditions. They have also introgressed many disease resistance genes from wild tomato relatives. Fruit quality has been improved mainly in the areas of shelf life, fruit homogeneity, shape and colour. However, consumers have frequently complained about tomato flavour over many years.

Improving taste by breeding is complex for several reasons. To begin with, sensory quality is a composite trait involving many components. Sugars and acids (responsible for sweet and sour flavours), as well as aroma (involving several volatiles) and texture (linked to firmness, meltiness, mealiness), contribute to flavour perception. Furthermore, the measurement of these compounds may be difficult; some are measured by only sensory analyses, but most of the components can be related to the chemical composition of the fruit. Tens of volatile compounds have been identified, but the list of those that are important for tomato aroma is very limited (Baldwin et al., 2000; Klee and Tieman, 2013).

http://dx.doi.org/10.19103/AS.2016.0007.13
© Burleigh Dodds Science Publishing Limited, 2017. All rights reserved.

In addition, most taste components are strongly influenced by the environment during plant and fruit growth and development (Causse et al., 2003), by the harvest stage (often immature) but also by post-harvest conditions (Kader et al., 1978; Whitaker, 2008). Many actors affect and can damage the flavour of a variety. Some important quality traits are also negatively correlated, like yield and sugar content or fruit shelf life and meltiness (due to physiological and genetic origin). This limits the options available for improving one trait without reducing others. Finally, quality has a subjective component based on individual consumer preferences. Economic factors can be a brake to optimizing taste which is not always sufficiently valued to attract a premium price for better varieties (Bellec-Gauche et al., 2015).

Nevertheless, it is well recognized that genetics has a fundamental impact on tomato flavour. Advances in molecular markers, and more recently the availability of the tomato genome sequence (Tomato Genome Consortium, 2012), have paved the way towards a better understanding of the genetic factors involved in fruit quality.

To improve flavour, several questions have to be addressed: What are consumer expectations? What is the genetic diversity available for breeding? What is the genetic control of tomato flavour? How to efficiently breed flavour? How can recent advances in genetics and genomics allow more efficient breeding? How environment influences flavour components? Emerging results to these questions will be presented in this section.

2 Genetic diversity of tomato flavour and consumer expectations

The first requirement to improve a trait is availability of genetic variability. Characterization of the genetic diversity tomato accessions for fruit quality components has revealed a large variation among traditional varieties and among wild related species for many traits as reviewed by a number of authors (Davies and Hobson, 1981; Stevens, 1986; Dorais et al., 2001; Causse et al., 2011). Metabolome profiling assessed several panels of tomato varieties and identified a large range of variations for primary and secondary metabolites (Schauer et al., 2006; Sauvage et al., 2014) as well as for volatile compounds (Tikunov et al., 2005; Bartoshuk and Klee, 2013; Rambla et al., 2014). All the metabolites appear to be influenced by growth conditions, varieties, ripening stages and storage conditions (Klee and Tieman, 2013).

Consumer preferences according to genetic diversity have been subject to a few studies (Sinesio et al., 2009; Causse et al., 2010). In the framework of a large European project, Eusol, 806 consumers from three countries (The Netherlands, France, and Italy) were presented with a set of 16 varieties representing the diversity of fresh tomato offer in order to evaluate their preferences. In parallel, expert panels in each country built sensory profiles of

© Burleigh Dodds Science Publishing Limited, 2017. All rights reserved.

the varieties. Preference maps were then constructed in each country, revealing the structure of consumer preferences and allowing identification of the most important characteristics. Then, a global analysis revealed that preferences were quite homogeneous across countries. This study identified the overall flavour and firmness as the most important traits for improving tomato fruit quality. It showed that consumer preferences from different European countries, with different cultures and food practices, are segmented following similar patterns when projected onto a common referential plan. Moreover, the results clearly showed that diversification of taste and texture is required to satisfy all consumers' expectations as some consumers preferred firm tomatoes, while others preferred melting ones and were more or less demanding in terms of sweetness and flavour intensity. Detailed comparisons also showed the importance of the fruit appearance in consumer preference.

To study the inheritance of taste-related traits and the influence of growth conditions, Causse et al. (2003) analysed the genetic variation of quality attributes in 35 hybrids and their 13 parental lines, grown in two contrasted environments. The 13 parental lines had various origins (old traditional inbred cultivars, experimental lines bred in the 1980s and lines used as parents of modern hybrid varieties). Each experiment was grown in spring in soil-less glasshouse conditions and in summer in the open field or under unheated plastic tunnels, in order to estimate the overall influence of environmental conditions on quality traits. As fruit size influences the judgement of taste panels, two experiments were set up, one involving large fruits, the other small fruits from hybrids between cherry tomato lines and large fruited lines. Among the main results on the genetic control of quality traits one can retain the following points:

- Differences for sensory traits among genotypes may be related to differences in fruit composition (sweetness and sourness with sugars and acids), but texture traits are more difficult to relate to instrumental measurements.
- Consumers particularly liked hybrids between old and modern lines with intermediate firmness. The preference for hybrids between large and cherry tomatoes confirmed the major role of sweetness and acidity in preference, which appeared more important than texture traits. The results also showed the importance of texture in consumer preference; thus if a good flavour is obtained, a good texture is the second criteria needed, at least in large-fruit hybrids.
- Correlations between sensory profiles and fruit composition allowed identification of the major components to be selected.
- Most of the physico-chemical traits, flavour attributes and firm texture showed a simple additive inheritance, in contrast to the aroma and other texture traits.

© Burleigh Dodds Science Publishing Limited, 2017. All rights reserved.

Several mutations affecting fruit ripening and shelf life are described. The most widely used in tomato breeding is *rin* (ripening inhibitor), which, in the heterozygous state, enables fruits to be kept for a few weeks (Davies and Hobson, 1981). Long shelf life cultivars have invaded the tomato market, but in the 1990s, their quality, particularly their colour and flavour, had been criticized by consumers (Jones, 1986; McGlasson et al., 1987). In the previous experiment, Causse et al. (2003) had produced and compared seven pairs of nearly isogenic hybrids, with or without the *rin* mutation at the heterozygous level. The presence of the *rin* mutation reduced consumer preference. Differences were detected by sensory profiles, *rin* hybrids having fruits on average 17% less sweet, with a lower tomato aroma, a higher 'strange' aroma and more mealy fruits. Instrumental firmness and sugar content were no different. These results confirmed the negative influence of the *rin* mutation on consumer preference, but also indicated that when transferred into a hybrid with high flavour, the negative influence of the mutation is reduced. Selection could thus be carried out to obtain much sweeter and perfumed lines combined with shelf life in *rin* hybrids.

3 Genes and quantitative trait loci affecting flavour

Many mutations involved in fruit development and composition have been discovered and used for fruit quality breeding. Table 1 lists the major mutations identified, which directly or indirectly impact fruit quality. They may induce variation in fruit colour or aspect. Some mutations impacting plant architecture, like *sp*, which controls the determinate/indeterminate growth, are also known to impact fruit quality. Today, several populations of EMS mutants have been produced in a few genetic backgrounds (Okabe et al., 2011). They enlarge the range of variations available and allow the rapid discovery of the responsible genes (Austin et al., 2011).

Most tomato fruit quality traits are quantitatively inherited. Tomato was among the first crop for which molecular markers were used to dissect the genetic basis of quantitative traits into QTL (quantitative trait loci, Tanksley, 1993). Since then, many QTL controlling yield and fruit quality-related traits have been mapped (Paterson et al., 1988, 1990, 1991; Azanza et al., 1994; Goldman et al., 1995; Grandillo and Tanksley, 1996; Tanksley et al., 1996; Fulton et al., 1997, 2000, 2002; Bernacchi et al., 1998a,b; Chen et al., 1999; Doganlar et al., 2002; Frary et al., 2004; Eshed and Zamir, 1995; see Labate et al. (2007) for review). Due to the very low polymorphism revealed at the within species level, these studies were performed on interspecific progenies derived from crosses between wild tomato species and tomato inbreds. In most of the studies a few QTL explained a large fraction (20–50%) of the phenotypic variation, acting in concert with minor QTL that could not be detected. Most of the QTL act in an

© Burleigh Dodds Science Publishing Limited, 2017. All rights reserved.

additive manner, but dominant and overdominant QTL have been detected (Paterson et al., 1988, 1991; de Vicente et al., 1993; Semel et al., 2006). Epistasis (interaction among QTL) was rarely detected unless a specific experimental design is used (Eshed and Zamir, 1996; Causse et al., 2007).

3.1 QTL for fruit size and shape

Grandillo et al. (1999) summarized the results of QTL mapping for fruit weight obtained in 17 studies. Six QTL explained more than 20% of the phenotypic variation. A common set of 28 QTL could be identified that frequently segregated in at least two populations. Nevertheless, only QTL cloning and complementation permits determination of whether each consensus QTL location corresponds to a single gene. Nowadays, only two fruit weight QTLs have been cloned by a map-based cloning approach. The first fruit size QTL to be cloned, *fw2.2* (Frary et al., 2000), controls up to 30% of the fruit size variation. It corresponds to an unknown function gene, ORFX, which acts on cell number in carpels before anthesis where it is differentially expressed between large and small fruits. However, its precise function is still unclear. The wild-type allele of ORFX negatively regulates cell division. The second QTL cloned for fruit weight (*fw3.2*), corresponds to a cytochrome P450 (Chakrabarti et al., 2013).

Locule number is another major component of fruit size and shape. Several QTL have been mapped (Lippman and Tanksley, 2001; van der Knaap and Tanksley, 2003; Barrero and Tanksley, 2004) for this trait. The two major QTL correspond to the mutations fasciated on chromosome 11 and *lc* on chromosome 2, with a strong epistatic interaction between these two genes (Lippman and Tanksley, 2001). Both mutations have been identified using a map-based cloning approach. The *lc* mutation is located near *Wuschel*, a gene that is responsible for stem cell fate in apical meristems, but 1500 bp upstream (Muños et al., 2011). Compared to *lc*, the fasciated mutation has a strong effect on locule number, increasing the trait from 3 to more than 6 locules. The QTL is located close to a Yabby-like transcription factor (Cong et al., 2008). Analysis of molecular diversity of the locus finally showed that the phenotype resulted from a large invertion (several kilobases) between the YABBY gene and the *CLV3* gene (Xu et al., 2015). Furthermore, Xu et al. (2015) underlined the role of Clavata pathway in interaction with arabinosyltransferase genes in meristem size and subsequently in fruit size.

For fruit shape, Grandillo et al. (1999) identified a set of 11 QTL from the six studies where the fruit length:diameter ratio was segregating. Three major QTL were identified, *ovate* on chromosome 2, *sun* on chromosome 7 and *fs8.1* on chromosome 8 (van der Knaap et al., 2002). The gene *ovate* encodes a predicted 40.7 kDa protein with an unknown function Liu et al. (2002). A mutation in the second exon of the ORF leads to a premature stop codon in the protein sequence. Plants containing this truncated protein exhibit the ovate

© Burleigh Dodds Science Publishing Limited, 2017. All rights reserved.

phenotype. The gene is expressed at the early developmental stages in flowers and fruits. Another mutation responsible for fruit length, *sun*, has been cloned (Xiao et al., 2008). The gene responsible for *sun* encodes an IQD protein. IQD proteins are found in plants and contain an IQ67 motif, which corresponds to a 67 amino acid motif. The function of this protein family is unknown, except for AtIQD1, which plays a role in the regulation of cytochrome P450 genes. The *sun* locus results from a retrotransposon duplication event. Functionally, the *sun* phenotype is due to a difference in the IQD gene expression. In wild-type plants, the gene is less expressed. The differential expression pattern could result from the new genomic context of the gene after duplication.

Rodriguez et al. (2011) showed that the combination of *lc*, *fas*, *sun* and *ovate* allows the classification of most shapes of the tomato fruit. Nevertheless, some QTLs modifying the effect of these genes remain to be detected. Fruit shape and size phenotypes are well described in thousands of natural accessions. The challenge is now to identify the molecular nature of QTLs with a weaker effect. Fruit shape and size are directly linked to developmental processes. To understand heterochrony, it is essential to characterize the natural diversity at the cellular level. For this purpose, new high-throughput tools have to be developed. Combining histological and molecular genetic regulation studies will help to clarify the precise mechanisms leading from stem cells to developed tomato fruits (Xu et al., 2015). Cell division and cell growth are two important mechanisms in fruit size; the two phases are distinct during fruit development. Consequently, genes involved in flower meristem development can be used as candidate genes for fruit size or shape (Barrero et al., 2006; Bauchet et al., 2014).

3.2 QTL for sugar and acid content

The review of Labate et al. (2007) summarizes the chromosome regions carrying QTL for sugar content or related traits (Soluble Solid Content-SSC, fructose, glucose or sucrose content) on the basis of 14 populations involving 8 different species (Paterson et al., 1988, 1990, 1991; Goldman et al., 1995; Azanza et al., 1994; Bernacchi et al., 1998a; Fulton et al., 1997, 2000, 2002a; Tanksley et al., 1996; Doganlar et al., 2002; Grandillo and Tanksley, 1996; Chen et al., 1999; Eshed and Zamir, 1995; Causse et al., 2004; Frary et al., 2004; Saliba-Colombani et al., 2001). A total of 95 QTLs were detected in 56 chromosomal regions. For the majority of QTL, the wild species alleles increased the sugar content. In 28 regions, QTL were detected in more than one population, and may possibly correspond to the same QTL. The large number of regions involved suggests that many mechanisms are responsible in increasing fruit sugar content. The same results were obtained for acid content (Fulton et al., 2002a; Causse et al., 2002, 2004), with only a few regions common to acid and sugar content. In contrast, frequent colocations between QTL for sugar content and fruit weight

© Burleigh Dodds Science Publishing Limited, 2017. All rights reserved.

(Grandillo et al., 1999) with opposite allelic effects were detected, suggesting pleiotropic effects of some common QTL. Few studies have reported QTLs for SSC with no apparent effect on fruit size (Yousef and Juvik, 2001; Fridman et al., 2004) and such antagonism may be responsible for the difficulty in simultaneously increasing fruit size and sugar content (Prudent et al., 2009). Part of this relationship is due to a dilution effect, but another part may be due to gene linkage as shown by fine mapping results (Lecomte et al., 2004).

The first QTL controlling SSC variation has been identified in a series of introgression lines derived from *S. pennellii* in an *S. lycopersicum* background (Eshed and Zamir, 1995). The QTL has been delimited to a region encompassing *Lin5* (Fridman et al., 2000), a gene encoding an apoplastic invertase expressed exclusively in fruits and flowers (Godt and Roitsch, 1997; Fridman and Zamir, 2003). Fridman et al. (2004) revealed that the wild species allele of *Lin5* was more efficient than the cultivated allele, due to a single-nucleotide substitution that coded for an amino acid residue close to the fructosyl-binding site of the enzyme. *In planta*, proof of the importance of *Lin5* in the control of the total soluble solids content in tomato has been confirmed by RNAi approach (Zanor et al., 2009a). The sucrose accumulation in *S. chmielewskii* and *S. habrochaites* fruits is associated with low-level acid invertase activity (Yelle et al., 1988, 1991; Chetelat et al., 1993).

Starch accumulates at the early stages of tomato fruit development, contributing approximately 20% of the dry weight of the fruit tissue at peak concentration, prior to the mature green stage. This starch is completely degraded in the ripe fruit, serving as a reservoir contributing to the soluble solids content (Dinar and Stevens, 1981; Ho, 1996). ADP-Glc pyrophosphorylase (AGPase) catalyses the synthesis of ADP-Glc, and is considered the first committed step in starch synthesis. Tomato plants (*S. lycopersicum*) harbouring the allele for the AGPase large subunit (AgpL1) derived from the wild species *S. habrochaites* (AgpL1 (H)) are characterized by higher AGPase activity and increased starch content in the immature fruit, as well as higher soluble solids in the mature fruit following the breakdown of the transient starch, as compared to fruits from plants harbouring the cultivated tomato allele (AgpL1(E), Schaffer et al., 2000). The increased activity of the *AgpL1H* in tomato fruit is due to an extended period of *AgpL1H* gene expression and subsequent stability of the S1-L1 heterotetramer (Petreikov et al., 2006). Similarly, the small subunit of ADP-glucose pyrophosphorylase (AGPase SS on chromosome 7) colocalized with QTLs for reducing sugars and fructose content (Causse et al., 2004).

3.3 QTL for volatile compounds

Volatiles are derived from the degradation of amino acids, fatty acids, carotenoids or phenolic compounds. A large range of variations for individual volatiles have been shown in panels of accessions (Klee and Tieman, 2013).

© Burleigh Dodds Science Publishing Limited, 2017. All rights reserved.

Table 1 Cloned genes with a phenotype related to fruit quality, plant, leaf or truss architecture; location on the tomato genome assembly

ITAG gene model	Gene symbol	Locus name	Chromosome	Start position	Phenotypic descriptors	References
Solyc01g079620	*y*	Colourless epidermis	1	71 255 600	Pink epidermis	Ballester et al. (2010)
Solyc10g081470	*L-2*	Lutescent-2	10	61858478	Altered chloroplast development and delayed ripening	Barry et al. (2012)
Solyc08g080090	*Gr*	Green flesh	8	60 582 066	Green fruit flesh	Barry et al. (2008)
Solyc06g074910	*C*	Potato leaf	6	42 804 036	Simple leaves	Busch et al. (2011)
Solyc03g031860	*r*	Phytoene synthase 1	3	8 606 749	Yellow fruit	Fray and Grieson (1993)
Solyc04g082520	*cwp1*	Cuticular water permeability 1	4	63 765 366	Microfissure/dehydration of fruits	Hovav et al. (2007)
Solyc10g081650	*t*	Carotenoid isomerase	10	62 006 972	Orange fruit flesh	Isaacson et al. (2002)
Solyc02g077390	*s*	Compound inflorescence	2	36 913 957	Inflorescence branching	Lippman et al. (2008)
Solyc02g077920	*Cnr*	Colourless non-ripening	2	37 323 107	Inhibition of ripening	Manning et al. (2006)
Solyc11g010570	*j*	Jointless	11	3 640 857	No pedicel abscission zone	Mao et al. (2000)
Solyc03g118160	*fa*	Falsiflora	3	61 162 449	Leafy inflorescence	Molinero-Rosales et al. (1999)
Solyc03g063100	*sft*	Single flower truss	3	30 564 833	Single flower truss	Molinero-Rosales et al. (2004)
Solyc01g056340	*hp-2*	De-etiolated 1	1	46 495 644	High pigment	Mustilli et al. (1999)

© Burleigh Dodds Science Publishing Limited, 2017. All rights reserved.

Solyc06g074350	*sp*	Self-pruning	6	42 361 623	Determinate plant habit	Pnueli et al. (1998)
Solyc10g008160	*u*	*Uniform ripening*	10	2 293 088	Increased chlorophyl content	Powell et al. (2012)
Solyc12g008980	*Del*	Delta	12	2 285 372	Orange fruit	Ronen et al. (1999)
Solyc06g074240	*B*	Beta-carotene	6	42 288 127	Increased fruit beta-carotene	Ronen et al. (2000)
Solyc03g083910	*sucr*	Sucrose accumulator	3	47 401 871	Accumulates predominantly sucrose in mature fruit, rather than glucose and fructose	Sato et al. (1993)
Solyc07g066250	*ls*	Lateral suppressor	7	64 958 148	Few or no axillary branches; corolla suppressed; partially male sterile	Schumacher et al. (1999)
Solyc02g090890	*hp-3*	Zeaxanthin epoxidase	2	46 947 557	High pigment in fruits	Thompson et al. (2000)
Solyc05g012020	*rin*	Ripening inhibitor	5	5 217 073	Never ripening	Vrebalov et al. (2002)
Solyc05g012020	*mc*	Macrocalyx	5	5 217 073	Large sepals	Vrebalov et al. (2002)
Solyc05g053410	*phyB2*	Apophytochrome B2	5	62 648 223	Red light reception	Weller et al. (2001)
Solyc10g044670	*phyA*	Apophytochrome A	10	22 854 459	Far-red light insensitive	Weller et al. (2001)
Solyc09g075440	*Nr*	Never ripe	9	62 631 866	Not ripening	Wilkinson et al. (1995)
Solyc04g076850	*e*	Entire leaf	4	59 354 677	Reduced leaf complexity	Zhang et al. (2007)

© Burleigh Dodds Science Publishing Limited, 2017. All rights reserved.

QTL for volatile compounds have been mapped in three populations. Saliba-Colombani et al. (2001) detected QTL for 12 volatile compounds among 18 that were quantified in the progeny of a cross involving a cherry tomato. Tieman et al. (2006) identified QTL for 23 volatiles in the population of introgression lines derived from *S. pennellii*. Twenty-five QTL were identified. Although ten volatiles were analysed in both studies, only three QTL were detected in the same regions, for phenylacetaldehyde on chromosome 8 (confirming the effect of the QTL *Malodorous*, named by Tadmor et al., 2002), on chromosome 9 for 2-methylbutanal and on chromosome 12 for pentanal. The content in some volatile compounds appeared strongly variable over years or environments (Tieman et al., 2006). This could partly explain the small number of QTL common to the two studies. In both studies, QTL for several volatiles were frequently in clusters. In a few cases these clusters corresponded to volatiles derived from the same metabolic pathway (related to fatty acid, carotenoid or amino acid degradation), suggesting the action of a gene within a single pathway. More frequently, colocalizations of QTL for volatiles derived from various metabolic pathways were shown, suggesting the presence of regulatory gene acting on several pathways. In *S. habrochaites* introgression lines, 30 QTL affecting the emission of one or more volatiles were mapped (Mathieu et al., 2009).

A few genes responsible for volatile accumulation have been identified (Table 2). The *ADH* gene coding for an alcohol dehydrogenase is involved in the ratio of hexanal to hexanol in the fruit (Speirs et al., 1998). *TomloxC*, a gene coding for a fruit-specific lipoxygenase has been shown to be related to the generation of volatile C6 aldehyde and alcohol compounds including

Table 2 Genes associated with volatile production in tomatoes

Gene	Associated volatile	Identification method	Reference
ADH	Hexanal:heanol ratio	BP	Speirs et al. (1998)
AADC	Phenylacetaldehyde, 2-phenylethanol, 1-nitro-2-phenethane, 2-phenylacetonitrile	BP/QTL	Tieman et al. (2006)
PAR	2-Phenylethanol	BP	Tieman et al. (2007)
LoxC	*Z*-3-Hexenal, *Z*-3-hexenol, hexanal, hexanol	CG	Chen et al. (2004)
SAMT	Methylsalicylate	BP	Tieman et al. (2010)
CTOMT	2-Methoxyphenol	BP	Mageroy et al. (2012)
CXE1	Multiple alcohols	QTL	Goulet et al. (2012)
CCD1	Multiple apocarotenoids	CG	Simkin et al. (2004)
GT1	Smoky aroma (phenylpropanoids)	QTL	Tikunov et al. (2013)

AADC, aromatic amino acid decarboxylase; PAR, phenylacetaldehyde reductase; LoxC, 13-lipoxygenase; SAMT, salicylic acid methyltransferase; CTOMT, catechol-*O*-methyltransferase; CXE1, carboxylesterase; CCD1, carotenoid cleavage dioxygenase. BP, Biochemical pathway; CG, candidate gene; QTL, positional cloning; Adapted from Klee et al. (2013).

© Burleigh Dodds Science Publishing Limited, 2017. All rights reserved.

hexanal, hexenal and hexenol (Chen et al., 2004). Two genes *LeAADC1* and *LEAADC2* are responsible for the decarboxylation of phenylalanine and subsequent synthesis of phenylethanol and related compounds (Tieman et al., 2006). The gene coding for the carotenoid cleavage dioxygenase 1 enzyme (*CCD1*) is involved in the synthesis of several aroma volatiles derived from carotenoid cleavage (Vogel et al., 2008). Tieman et al. (2007) showed that phenylacetaldehyde reductases (PAR) catalyse the last step in the synthesis of the aroma volatile 2-phenylethanol. A salicylic acid methyl transferase has been shown to be involved in the synthesis of methyl salicylate (Tieman et al., 2010). Tikunov et al. (2013) identified a mutation in a glycosyltransferase which is responsible for the release of smoky aroma related to phenylpropanoid compounds.

4 Tomato texture

Fruit texture is a complex breeding objective, as it involves the fruit firmness and shelf life, but also refers to a wider range of sensory attributes such as crispiness, juiciness, meltiness or mealiness. Texture is dependent on the overall fruit structure and spatial organization, the cellular morphology of main tissues, the cell turgor in addition to the biochemical and mechanical properties of the cell walls (Shackel et al., 1991; Harker et al., 1997; Chaib et al., 2007). In fleshy fruits, texture does not only influence the purchasing power of the consumer and consumer acceptance, but it also has a significant impact on overall organoleptic quality, shelf life, and transportability (Seymour et al., 2002) and it strongly interferes with the perception of flavour and aroma (Causse et al., 2003, 2011). After harvest, texture evolves rapidly, while membrane and cell wall breakdown occurs in relation to turgor loss and to enzyme-orchestrated cell wall loosening. Internal hormonal stimulation, as well as environmental factors such as light, temperature, water and nutrient supply, regulates ripening. Fruit texture is, thus, essentially, an unstable characteristic closely related to shelf life (Seymour et al., 2013).

Fruit firmness has been studied in several quantitative genetic studies. Labate et al. (2007) present a summary of QTL controlling fruit firmness in nine populations (Bernacchi et al., 1998a; Causse et al., 2002; Doganlar et al., 2002; Frary et al., 2003, 2004; Fulton et al., 1997, 2000; Tanksley et al., 1996). Forty-six QTL controlling firmness were mapped using seven different populations. More than half of the QTL were grouped in clusters of three to four QTL. These clusters were localized on chromosomes 1, 2, 4, 5, 9, 10 and 11. Chapman et al. (2012) dissected a fruit firmness QTL on chromosome 2 and revealed a complex locus with epistatic interactions.

Our current understanding of ripening mechanisms and the molecular basis of fruit texture in fleshy fruit mainly relies on transgenic or mutant plant analysis (Giovannoni, 2007; Seymour et al., 2002; Vicente et al., 2007).

© Burleigh Dodds Science Publishing Limited, 2017. All rights reserved.

The pleiotropic *rin* (ripening inhibitor) recessive mutation blocks the ripening process. Mutant fruits fail to produce ethylene and are unable to ripen under ethylene treatment, although they are responsive to ethylene. Breeders have extensively used the *rin* mutation and hybrids (*rin/Rin*) form the basis for most present-day production of slow ripening, long shelf life, fresh market tomatoes. The gene underlying the mutant was cloned; it encodes a partially deleted MADS-box protein of the SEPALATTA clade (Vrebalov et al., 2002; Ito et al., 2008; Hileman et al., 2006). Another mutation, *Cnr* (colourless non-ripening) corresponds to an epigenetic mutation in a member of the same gene family (Manning et al., 2006).

The decrease in tomato firmness coincides with the dissolution of the cell wall middle lamella, resulting in lower intercellular adhesion, depolymerization and solubilization of pectic and hemicellulosic cell wall polysaccharides (Brummell and Harpster, 2001; Rose et al., 2004). Although many genes have been identified, their role in the natural variation of fruit texture has been rarely demonstrated. Polygalacturonase and pectin methylesterase were long considered as major enzymes for pectin depolymerization and de-esterification, but antisense mRNA-mediated suppression had only a minor effect on cell wall loosening and failed to reduce fruit softening (Tieman and Handa, 2004; Brummell and Harpster, 2001). Similarly, weak effects on fruit texture were obtained with several other ripening-related cell wall-modifying enzymes (reviewed by Rose et al., 2003). A multigene family of 7 members encodes β-galactosidases, whose activity increases during ripening. Tomato fruits with suppressed β-galactosidase expression soften more slowly during ripening (Smith et al., 2002), demonstrating that pectic side chains contribute to fruit texture. The possible involvement of pectin lyase and acetylesterases in pectin breakdown is proposed by Vicente et al. (2007).

During ripening xyloglucan depolymerization occurs within the hemicellulose fraction without clear identification of the responsible proteins. Some experimental clues indicate that endoglucanases and endo transglucosylases might catalyse xyloglucan degradation, but this possibility has to be further explored (Vicente et al., 2007; Saladie et al., 2006).

Expansins are proteins that contribute to cell expansion by a non-enzymatic cell wall-loosening biomechanical process (Cosgrove, 1998). They are present and active in ripening tomato fruit (Rose et al., 2000) where they participate in cell wall disassembly and may enhance cellulose degradation by cellulases. Their exact contribution to fruit softening is yet to be demonstrated.

The difficulty encountered in identifying one or a few key determinants of fruit softening is due to the complexity of the process probably involving many different cell wall actors in a fine orchestrated manner. Meli et al. (2009) suggested that N-glycoprotein-modifying enzymes such as α-mannosidase and β-D-acetylhexosaminidase may play a role in tomato ripening-associated

© Burleigh Dodds Science Publishing Limited, 2017. All rights reserved.

fruit softening. RNAi downregulation of these two ethylene-regulated genes led to subsequent downregulation of many genes that are associated with tomato ripening and cell wall degradation. Moreover, identification of new cell wall-modifying enzymes in order to gain new insight into the biochemical processes underlying fruit ripening is the present-day challenge; proteomic studies represent a promising perspective in this area because of the high level of post-transcriptional regulation of cell wall proteins (Minic et al., 2009).

Finally, tomato fruit texture and shelf life may rely on other physiological mechanisms unrelated to cell wall loosening (Matas et al., 2009). In particular, fruit water status and cuticle structure may be important factors related to shelf life. The Delayed Fruit Deterioration (DFD) tomato cultivar which is able to remain firm for several months, exhibits normal ripening and cell wall loosening but very low fruit transpiration, high cellular turgor and a different cutin composition as compared to a control cultivar (Saladie et al., 2007). Such results, therefore, emphasize the possibility of a disconnection between pericarp firmness and fruit shelf life. Moreover, fruit firmness at harvest may be disconnected from its ability to remain firm after a period of storage especially at cold temperature. Recently, Page et al. (2010) compared the behaviour of two isogenic lines for a firmness QTL at harvest and during cold storage and found that the line possessing the favourable allele for firmness had the lowest storage ability. The lack of ability to remain firm was correlated with the lower expression of genotype-specific protective proteins, among others, heat shock proteins. Ascorbic acid redox state has also been shown to be involved in fruit shelf life and tolerance to cold storage (Stevens et al., 2008).

Fruit texture and shelf life capacity, and potentially, the interaction with susceptibility to pathogens (Cantu et al., 2008) represent a highly challenging research area which is currently benefiting from several genomic approaches.

5 New approaches to tomato flavour diversity and genetic control

During the last decade, the advent of the high-throughput sequencing and genotyping technologies enabled the collection of data at the genome-wide scale for hundreds of thousands of single-nucleotide polymorphisms (SNPs) at a reasonable cost. This task was facilitated by the release of the reference genome of major crops, among which was the tomato genome in 2012 (TGC, 2012). Then, tools such as the SolCAP SNP genotyping array were derived in tomato (Hamilton et al., 2012; Sim et al., 2012). On the basis of 7720 SNP markers, this array was largely used for different purposes, including the investigation of the tomato worldwide germplasm nucleotide diversity (Blanca et al., 2012; 2015), the study of the linkage disequilibrium decay along chromosomes (Sim et al., 2012a) or the establishment of reference linkage maps (Sim et al., 2012b)

© Burleigh Dodds Science Publishing Limited, 2017. All rights reserved.

to pave the way for the mapping of quantitative traits linked to agronomical traits. In parallel, the phenotyping of multiple traits related to fruit quality in multiple environments for large populations obtained from bi-parental or multi-parental crosses was achieved to decipher the genetic basis (i.e. broad sense of heritability, number of loci) of these traits and their interaction with the environmental conditions.

To overcome the main limitation of the QTL experimental design (essentially the lack of recombination), benefit was taken from the ancestral polymorphism found in natural population or germplasm core collections to identify the underlying molecular determinants of agronomical traits. In 2006, a mixed linear model (MLM), based on the statistical model described in Henderson (1975), was proposed to test the statistical link between the genotypic and the phenotypic data in a collection of maize varieties (Yu et al., 2006), while taking into account the confounding effect of the pairwise genetic relatedness (the so-called K-matrix) between accessions and population structure (the Q-matrix). The genome-wide association (GWA) approach was adopted in many crops, including the tomato. Ranc et al. (2008) initiated the building of a reference core collection of 360 accessions on the basis of the genetic diversity revealed at 20 microsatellite markers. Despite being restricted to a single chromosome, a proof of concept was presented with the mapping of genotype-phenotype associations for a flavour trait (solid soluble content) in this core collection. Subsequent studies applied the approach at the genome-wide scale while detecting more and more loci. Using a set of 192 SNP markers genotyped in 188 accessions, Xu et al. (2012) identified 2, 16 and 17 loci associated to titrable acidity, soluble solids and sugar contents of the fruit, while the phenotypic heritabilities were estimated to 0.75, 0.73 and 0.63, respectively. A similar study, in terms of experimental design, with 174 accessions (both *S. lycopersicum* and *S. l.* var *cerasiforme*) and 182 SSR, identified 17 and 22 associated loci for fruit weight and ascorbic acid content, supporting the polygenic genetic architecture of these traits (Zhang et al., 2016). Favourable allelic combination between loci associated to fruit quality, such as pH, titrable acidity or SSC, were defined from a classical (K+Q) MLM in the core collection of 96 fresh market and processing tomatoes (Ruggieri et al., 2014).

To deepen the framework of GWA, extensive work towards developing SNP was also achieved through resequencing projects, especially for larger collections of tomato accessions. For example, Yamamoto et al. (2016) identified over 50,000 markers that were implemented in a GWA approach for traits of agronomical interest (i.e. plant height, fruit size), including traits related to flavour such as soluble solid content. By using a multi-locus mixed model (see Segura et al., 2012), Sauvage et al. (2014) provided an extended list of loci notably associated to important metabolic compounds for flavour such as fructose, SSC and malic and citric acids. In this experimental design, the broad

© Burleigh Dodds Science Publishing Limited, 2017. All rights reserved.

sense of heritability was estimated to 0.56, 0.6, 0.64 and 0.42, respectively, demonstrating that not all the genetic variabilities of these traits have been captured by the molecular markers. In addition, the genotype by environment interaction may certainly bias these heritability estimates as flavour traits are under the influence of growing conditions.

In recent years, the high-throughput genomic produced large amounts of SNP genotyping data that are now overcome by large resequencing projects not only in the cultivated tomato (see Aflitos et al., 2014 and Lin et al., 2014) but also in its wild relatives (i.e. *Solanum pennellii*, see Bolger et al., 2014), from which fragments carrying genes of interest were introgressed. The statistical approaches were refined, notably to take into account confounding effects such as genetic relatedness or population stratification to improve the power of the GWA approach (see Tucker et al., 2014). One of the main conclusions of these GWA studies is the polygenic architecture of the traits related to flavour in tomato that may explain why breeding for improving such traits remains a challenging endeavour. Another conclusion is the intraspecific genetic variability that might be still exploited for enhancing tomato fruit quality, especially in the *S.l.* var *cerasiforme* group. While being genetically diverse, this group has the advantages to be in admixture between the closest wild relative tomato (*S. pimpinellifolium*) and the big-fruited cultivated tomato (*S. lycopersicum*) and having a large phenotypic diversity for many quality components. Thus, focusing on this group might improve the power of GWA and remove the population structure confounding effect, especially when the quality traits are correlated to the population structure (i.e. fruit weight or sugar content). In addition to cheery-type tomato, landraces remain underexploited and might also provide a viable indirect selection tool in future practical breeding programmes (Sacco et al., 2015).

New types of populations involving several parental lines like MAGIC (Multi-allelic Genetic Intercross) have also been found useful to map QTL into small confidence intervals. Combined with the resequencing of the parental lines, a direct access to putative polymorphisms under the QTL could be proposed (Pascual et al., 2015). The comparison of biparental, MAGIC and GWAS panels confirmed the complementarity of the three kinds of populations (Pascual et al., 2016).

On the basis of these sets of associated molecular markers, two different steps forward have to be achieved. The first one relies on the dissection of the molecular mechanisms underlying flavour traits. Here again, to reach this objective, the broad panel of biotechnologies (the so-called 'new breeding technologies') offered is plethora: RNAi, TALEN or CRISPR\Cas9 can be implemented to validate functionally the most promising loci identified by the GWA. The second step forward aims at implementing the molecular information gathered into a marker-assisted selection (MAS) breeding scheme and sustain

© Burleigh Dodds Science Publishing Limited, 2017. All rights reserved.

the varietal innovation. For traits governed by a low to moderate number of genes (,10), this approach is feasible. However, for complex and polygenic traits, such as the ones determining flavour in tomato, the MAS approach is limiting in this case.

6 From MAS to genomic selection for flavour breeding

Candidate loci affecting traits of agronomical interest are plentiful, but very few markers have been exploited so far, compared to the numerous linkage or association studies published (Jonas and de Koning, 2013). One of the main reasons is the variation of the marker effects between environments and populations leading to non-consistent results (Bernardo et al., 2008). MAS is still successful in assisting breeding but remains limited to a moderate number of markers. Thus, MAS is now being extended by the latest selective breeding approach, the genomic selection (GS), a multiple markers and genome-wide scale approach.

In the early 2000s, MAS for quality traits was initiated for five QTLs controlling fruit quality traits in tomato. This investigation revealed epistatic interactions between the QTLs and the genetic background, limiting the breeding efficiency according to the recipient parent (Lecomte et al., 2004). While being efficient in improving quality traits, the introgression of large chromosomic regions favoured the linkage drag of undesired alleles. This study demonstrated how challenging is the MAS for complex traits and how many generations of crosses will be required to clean the genetic background from the linkage drag, especially for multiple traits.

Facing this limitation, plant breeders are now evaluating the potential of the latest selective breeding strategy, the GS. The funder paper published by Meuwissen et al. (2001) describes this approach that is aimed at estimating the breeding value of an individual from the genome-wide genetic information. Precisely, an effect is attributed to all the markers identified in the genome of individuals, from which genotype and phenotype are known. These individuals compose the so-called training population (TP). Then, the sum of all these effects is the genomic estimate of the breeding value (GEBV) of each individual, for the considered trait: the larger, the more interesting is the individual to reproduce. Within the cross-validation step of the GS process, statistical models are tested for their accuracy to predict the phenotype of the individuals of the TP, on the only basis of the genotypic information. One or several may be accurate when the coefficient of correlation between the observed phenotype and the predicted one is high to very high (0.5–0.9). Then, the accurate prediction model is implemented in a real breeding scheme. GS has been largely investigated and implemented in animals, especially for dairy cattle, for which the evaluation of the real genetic gain has been published (Patri et al., 2011), supporting the

© Burleigh Dodds Science Publishing Limited, 2017. All rights reserved.

success of the approach. However, transferring the methods and knowledge obtained from animal to plant breeding is not trivial as the animal model does not take into account biological parameters like genotype by environment interactions.

In plant breeding, the GS potential is now being tested. More advanced experiments have been conducted in the annual crop wheat and maize (see Poland et al., 2012 and Bernardo and Yu, 2007). The potential of GS is also being tested in perennial species such as Spruce (Beaulieu et al., 2014), wine grape (Fodor et al., 2014) or apple (Kumar et al., 2012; Muranty et al., 2015). In tomato, the amount of data produced for the GWA experiments was further used to test the potential of GS to improve quality traits. The published studies rely on either the combination of GWA, GS cross validation and recurrent GS simulation (Yamamoto et al., 2016) and on the estimation of the effects of various parameters (i.e. marker density, size and composition of the TP) on the accuracy of the statistical models (Duangjit et al., 2016). In both studies, the targeted traits are mainly related to quality traits, especially flavour with sugar content or acidity. The accuracy of the GEBV in the first study was evaluated from a set of ~16k SNP in a TP of 96 accessions, while for the second study, the accuracy was evaluated from ~7k SNP in a TP of 122 accessions. For the common phenotypes to both studies, the estimated accuracy was similar with 0.807 and 0.714 for SCC, for example. This demonstrated the reliability of the approach. However, these results have to be carefully interpreted, as in both studies, phenotype values were estimated across several years and growing environment, controlling for the G 3 E interactions. Briefly, both studies demonstrated that reliable phenotype predictions could be obtained in tomato for highly and moderately heritable traits, stimulating interest to implement this approach in a large-scale breeding scheme. However, GS is still in its infancy in tomato and other crops. Priority has first been given to cross validation, but GS has to take the next step by delivering more of its promises.

7 Interactions genotype by environment: a tool for breeding good tomatoes

Tomatoes are produced year-round under contrasting environmental conditions, triggering seasonal variations in their sensory quality. Over the tomato growing cycle, different factors such as light intensity, air and soil temperatures, plant fruit load, plant mineral nutrition or water availability influence the final fruit quality (reviewed in Davies and Hobson, 1981 and Poiroux-Gonord et al., 2010). Variations in temperature and irradiance during ripening affect carotene, ascorbic acid and phenolic compound content in the fruit, although acid and sugar content are not modified considerably by these two factors (Venter et al., 1977; Rosales et al., 2006 and Gautier et al.,

© Burleigh Dodds Science Publishing Limited, 2017. All rights reserved.

2008). Changes in plant fruit load through trust pruning modify fruit dry matter content and final fruit fresh weight by disrupting the carbon flux entering to the fruit (Bertin et al., 2000; Guichard et al., 2005). Water limitation and irrigation with saline water may impact positively tomato fruit quality, mainly through an increase in sugar content in fruit (either by concentration or by accumulation effect) and through contrasted effects on the secondary metabolite contents (Mitchell et al., 1991; De Pascale et al., 2001; Nuruddin et al., 2003; Johnstone et al., 2005; Gautier et al., 2009; Ripoll 2016a; Ripoll 2016b). The effects reported on fruit composition are associated or not to large yield loss depending upon the intensity and duration of the treatment and the development stage of the plant (see Ripoll et al., 2014 for review). They result from modifications of the water and carbon fluxes imported by the fruit during its growth (Guichard et al., 2001; Albacete et al., 2013; Osorio et al., 2014).

Thus, the optimization of the growing practice, in particular water management, is considered in horticultural production as a tool to manage fruit quality while limiting yield losses, offering the opportunity to address simultaneously environmental issues and consumer expectations of tastier fruits (Stikic et al., 2003; Fereres et al., 2006; Costa et al., 2007). The genetic variability of tomato response to water limitations and other abiotic constraints and their combination still need to be deciphered to develop genotypes adapted to these practices (Poiroux-Gonord et al., 2010; Ripoll et al., 2014). Large phenotypic variation in response to a wide range of climate and nutrition conditions exists in the genus *Solannum* at both inter- and intra-species levels (reviewed in Labate, 2007). The TGRC (Tomato Genetics Resource Center, UC Davis) maintains wild and cultivated accessions with known or inferred tolerances to various abiotic stresses, including drought, flooding, high temperature, chilling injury, aluminium toxicity, salinity and/or alkalinity, providing useful starting material for breeding, genetic mapping and other uses.

Several authors attempted to measure genotype by environment (G x E) interactions on tomato fruit quality by repeating a same experiment in different locations or/and under several growing facilities (Auerswald et al., 1999; Johansson et al., 1999; Causse et al., 2003) or by building experimental design to isolate the effect of particular environmental factors on a large number of genotypes (see Semel et al., 2007; Albert et al., 2016; Gur et al., 2011 for water availability and Monforte et al., 1996; Monforte et al., 1997a, Monforte et al., 1997b for salt stress). In the different experiments, the G x E interaction was significant for the fruit quality traits measured (including fruit fresh weight, secondary and primary metabolism contents and fruit firmness), but generally accounted for a low part of the total variation in comparison with the genotype main effect. Albert et al. (2016) dissected further the genotype by watering regime interaction in an intraspecific *S. lycopersicum* recombinant inbred line population grown under two contrasting watering regimes in two

© Burleigh Dodds Science Publishing Limited, 2017. All rights reserved.

locations. In their studies, the interaction resulted from genotype re-ranking across the watering regime rather than scale changes. Besides, they identified large genetic variation and genetic heritabilities under both watering regimes, encouraging the possibility of developing tomato genotypes with an improved fruit quality under deficit irrigation.

The emergence of high-throughput genomic tools and the availability of genome sequences facilitate the decomposition of the genotype by environment interactions into underlying QTLs and/or genes. For this purpose, the first approach consists in modelling the effect of QTLs across different environmental conditions. In tomato, a few QTL studies considering the interaction with environmental variables at the fruit level were reported. They mainly pertained to response to salt stress (Monforte et al., 1996; Monforte et al., 1997a; Monforte et al., 1997b; Uozumi et al., 2012; Asins et al., 2015) and drought stress (Gur et al., 2011; Albert et al., 2016). All these studies identified numerous loci with low to medium effect, suggesting a strongly polygenic architecture of tomato fruit response to environmental constraints. Gur et al. (2011) described drought-responsive QTLs for fruit fresh weight and sugar content mainly expressed by the shoot in a reciprocal-grafting experiment, whereas Monforte et al. (1997b) identified QTLs with changing additive and epistatic effects according to the salinity level of the watering solution. Nevertheless, the authors mostly compared QTLs at different map positions and with different effects across experiments and conditions, which may be questionable as these comparisons depend upon the mapping significance threshold. Besides, the populations used were mainly introgression lines involving wild relative species (*Solanum habrochaites, Solanum pennellii* and *Solanum pimpinellifolium*) and the confidence intervals obtained remain large and difficult to transpose into the cultivated tomato. Statistical improvements allowing to explicitly take into consideration the effect of environmental variables in mapping models are available in the linkage (Van Eeuwijk et al., 2010; El Soda et al., 2014; Li et al., 2015; Verbyla et al., 2014) and association (Korte et al., 2012; Saïdou et al., 2014) framework and permit to analyse more complex population designs (RILs, GWA collections, multi-allelic MAGIC populations). These models offer the possibility to test properly the QTL by environment interactions and to identify QTLs whose effects are changing according to the environment. Recently, by applying such models in their *S. lycopersicum* RIL population grown under two contrasted watering regimes, Albert et al. (2016) mapped a total of 56 QTLs for plant and fruit quality traits, among which 20% presented effects changing direction or intensity according to the irrigation treatment. This proportion of interactive QTL is roughly identical to the results obtained in other crop species (Tinker et al. (1996) in barley; Sari-Gorla et al. (1997) and Melchinger et al. (1998) in maize) and should be considered for crop improvement.

© Burleigh Dodds Science Publishing Limited, 2017. All rights reserved.

The second strategy to decipher the G x E interaction into QTLs consists in constructing composite variables measuring phenotypic plasticity to deal with univariate QTL mapping models. These variables can be simple ratio or difference between the values of a trait measured under two contrasted conditions or parameters derived from more or less complex crop models integrating multiple environments. The ecophysiological models constitute adequate tools for analysing the genotype by environment interactions since they integrate environmental and genetic effects on individual processes and are able to predict interactions among processes during fruit development (Bertin et al., 2010). The Virtual Fruit Model was developed by Fischman and Génard (1998) to describe both the water and dry matter accumulation rates in fleshy fruits. This model was powerful in assessing the impacts of fruit load in tomato (Prudent et al., 2011) or of water deficit on fruit growth in different species, among which are peach (Quilot et al., 2005), mango (Lechaudel et al., 2006), kiwifruit (Hall et al., 2013) and tomato (Liu et al., 2007). When plant traits are generally dependent on genotype, environment and cultural practices, model parameters are, ideally, independent of the environment and management and are amenable to QTL analysis within univariate mapping models. Then, plants carrying different combinations of QTL might be simulated and tested under different environmental constraints, helping to select the best ideotypes as illustrated by Reymond et al. (2003) in maize and Quilot et al. (2005) in peach.

The major lock to decipher the genetic determinants of fruit quality and its response to environmental constraints remains in the ability to phenotype large number of accessions (to maintain a satisfying power in the genetic analysis) under contrasted environmental conditions. This limitation tends to be bypassed by the development of high-throughput phenotyping platforms, allowing to phenotype a large number of plants in finely controlled environments. Nevertheless, these platforms are generally difficult to adapt to grow tomato up to the production of mature fruits and may differ strongly from the conditions suffered by the plants in real production conditions. The second type of high-throughput phenotypic platforms will characterize the 'invisible' phenotypes, such as secondary metabolites, that are major determinants of flavour (see Tikunov et al., 2013 for an example for aroma in tomato). Thus, a second wave of large phenotyping data may submerge the discipline of plant breeding. This will require new statistical models that handle these large datasets and the mixing with the other 'omics' data (i.e. transcriptomic profiling, RNAseq) to push the plant breeding into the era of system biology. Tomato has proven to be one of the best models for the integration of multiple levels of information to sustain breeding (Pascual et al., 2013). Finally, changes in gene expression were shown to be the key process of tomato response to environmental constraints (Chen and Tabaeizadeh 1992; Thompson et al., 1995; Zhou et al.,

© Burleigh Dodds Science Publishing Limited, 2017. All rights reserved.

2007). In the near future, measures of gene expression at whole genome scale (using RNAseq technologies or microfluidic qPCR or microarrays) in a large set of accessions under contrasted conditions associated to eQTL (expression QTL) mapping may help to decipher the molecular basis of tomato response to environmental constraints.

8 Future trends

At the genetic level, many studies focused on the genetic variation and the genes regulating the fruit quality components, but only a few QTL were finely characterized. Furthermore, the integration of these results into breeding process is still incomplete.

Today, many new genomic and genetic resources are available. Several tilling mutant collections were developed and constitute novel sources of variation (Okabe et al., 2011). A high-quality tomato genome sequence is publicly available together with a large set of transcriptome data in several accessions and for different organs, stages and conditions. The high-throughput sequencing technology allows rapid mapping and cloning of new genes. Thus in the coming years, polymorphisms and genes involved in tomato flavour should be identified and used for breeding better tomatoes. Combined with adapted growth conditions they should better answer consumer expectations.

9 Conclusion

Tomato fruit quality is a complex trait involving a number of components, including fruit aspect, flavour, aroma and texture. A large range of genetic diversities have been shown in tomato for fruit quality components. Although a few major mutations may have a huge effect on fruit quality (notably the *rin* mutation), most of the components have a quantitative inheritance. Several QTL mapping experiments have been performed, mostly on interspecific progeny. Many loci and QTL have thus been detected, revealing some QTL cluster regions. Tomato is a model plant for fruit development and composition, and knowledge about its physiology rapidly increases and several genes affecting fruit quality are discovered. New approaches such as genome-wide association studies or MAGIC populations using the genome information allow a higher precision of QTL location.

Environment and post-harvest conditions may also strongly affect fruit quality and interact with the genotype limiting the genetic progress. More research in this field is thus necessary to identify the processes affected by the environment and assay whether light stress can improve sensory quality.

© Burleigh Dodds Science Publishing Limited, 2017. All rights reserved.

Although many QTL studies have been performed, MAS has been rarely set up and results are mitigated. Particularly the negative correlation between fruit size and sugar content has limited genetic progress. Today, new hopes arise from GS, although the impact of such method still needs to be demonstrated.

10 Where to look for further information

Introductions to the subject for non-specialists

Book chapters:

- Quilot-Turion, B. and Causse, M. (2014). Natural diversity and genetic control of fruit sensory quality. In Pravendra, N., Bouzayen, M., Mattoo, A. K. and Pech, J. C. (Eds), Fruit Ripening: Physiology, Signalling and Genomics, pp. 228–45. CABI.
- Causse, M. (2008). Genetic background of flavour: the case of the tomato. In Brückner, B. and Wyllie, S. G. (Ed.), Fruit and Vegetable Flavour. Recent Advances and Future Prospects, pp. 229–53. CRC Press, Boca Raton.
- Labate, J. A., Grandillo, S., Fulton, T., Muños, S., Caicedo, A. L., Peralta, I., et al. (2007). 1. Tomato. In C. Kole (Ed.), Genome Mapping and Molecular Breeding in Plants, Vol. 5, Vegetables, pp. 11–135. Springer-Verlag Berlin Heidelberg.

Any seminal articles or books which have shaped the subject:

- Causse, M., Damidaux, R. and Rousselle, P. (2007). Traditional and enhanced breeding for fruit quality traits in tomato. In Razdan, M. K. and Mattoo, A. K. (Eds), Genetic Improvement of Solanaceous Crops, Vol. 2, p. 637. Science Publishers, Enfield, USA.
- Klee, H. J. and Tieman, D. M. (2013). Genetic challenges of flavor improvement in tomato. Trends Genet 29, 257–62.
- Bartoshuk, L. M. and Klee, H. J. (2013). Better fruits and vegetables through sensory analysis. Current Biology 23, R374-R378.
- Causse, M., Saliba-Colombani, V., Lecomte, L., Duffé, P., Rousselle, P. and Buret, M. (2002). Genetic analysis of fruit quality attributes in fresh market tomato. Journal of Experimental Botany 53/377, 2089–98.

Any key websites worth visiting to keep up to date with trends:

- Solgenomics (https://solgenomics.net)

Any key journals or conferences:

- Solanaceae Genome Congress, Tomato Eucarpia Meetings, Tomato Roundtable Meetings; Solanaceae Workshop – Plant and Animal Genomes Meeting

© Burleigh Dodds Science Publishing Limited, 2017. All rights reserved.

Any major international research projects:

- Tomato Genome cooperative for sequencing the tomato genome (https://solgenomics.net/); FP6 EUSOL (https://www.eu-sol.wur.nl/), H2020 Traditom (http://traditom.eu/)

Top five or more research centres that readers can investigate for possible collaboration as well as to keep up with research trends:

- USA (Jim Giovanonni, Harry Klee), France-INRA (Christophe Rothan, Mondher Bouzayen), Netherlands-WUR (Arnaud Bovy), Spain-CSIC-Valencia (Antonio Granell).

11 References

Aflitos, S., Schijlen, E., de Jong, H., et al. (2014). Exploring genetic variation in the tomato (*Solanum* section *Lycopersicon*) clade by whole-genome sequencing. *The Plant Journal* 80(1): 136–48.

Albacete, A. A., Martínez-Andújar, C. and Pérez-Alfocea, F. (2014). Hormonal and metabolic regulation of source-sink relations under salinity and drought: From plant survival to crop yield stability. *Biotechnology Advances*, 32(1), 12–30.

Albert, E., Gricourt, J., Bertin, N., Bonnefoi, J., Pateyron, S., Tamby, J. P. and Causse, M. (2016). Genotype by watering regime interaction in cultivated tomato: lessons from linkage mapping and gene expression. *Theoretical and Applied Genetics* 129(2), 395–418.

Asins, M. J., Raga, V., Roca, D., Belver, A. and Carbonell, E. A. (2015). Genetic dissection of tomato rootstock effects on scion traits under moderate salinity. *Theoretical and Applied Genetics* 128(4), 667–79.

Auerswald, H., Peters, P., Bruckner, B., et al. (1999). Sensory analysis and instrumental measurements of short-term stored tomatoes (*Lycopersicon esculentum* Mill.). *Postharvest Biology and Technology* 15, 323–34.

Austin, R. S., Vidaurre, D., Stamatiou, G., Breit, R., Provart, N. J., Bonetta, D., Zhang, J., Fung, P., Gong, Y., Wang, P. W., McCourt, P. and Guttman D.S. (2011). Next-generation mapping of Arabidopsis genes. *The Plant Journal* 67, 715–25.

Azanza, F., Young, T. E., Kim, D., Tanksley, S. D. and Juvik J. A. (1994). Characterization of the effects of introgressed segments of chromosome 7 and 10 from *Lycopersicon chmielewskii* on tomato soluble solids, pH and yield. *Theoretical and Applied Genetics* 87, 965–72.

Baldwin, E., Scott, J., Shewmaker, C. and Schuch, W. (2000). Flavor trivia and tomato aroma: biochemistry and possible mechanisms for control of important aroma components. *Hortscience* 35, 1013–22.

Ballester, A. R., Molthoff, J., de Vos, R., et al. (2010). Biochemical and molecular analysis of pink tomatoes: deregulated expression of the gene encoding transcription factor SlMYB12 leads to pink tomato fruit color. Plant Phys 152: 71–84

Barrero, L. S., Cong, B., Wu, F., et al. (2006). Developmental characterization of the fasciated locus and mapping of Arabidopsis candidate genes involved in the control of floral meristem size and carpel number in tomato. Genome 49, 991–1006.

© Burleigh Dodds Science Publishing Limited, 2017. All rights reserved.

Barrero, L. S., Tanksley S.D. (2004). Evaluating the genetic basis of multiple-locule fruit in a broad cross section of tomato cultivars. *Theoretical and Applied Genetics* 109:669–79

Barry, C. S., Aldridge, G. M., Herzog, G., et al. (2012). Altered chloroplast development and delayed fruit ripening caused by mutations in a zinc metalloprotease at the lutescent2 locus of tomato. *Plant Physiology* . 159(3): 1086–98

Barry, C. S., McQuinn, R. P., Chung M.Y., et al. (2008). Amino acid substitutions in homologs of the STAY-GREEN protein are responsible for the green-flesh and chlorophyll retainer mutations of tomato and pepper. *Plant Physiology* 147(1), 179–87.

Bartoshuk, L. M. and Klee H.J. (2013). Better fruits and vegetables through sensory analysis. *Current Biology* 23, R374–8.

Bassel, G. W., Mullen, R. T. and Bewley J.D. (2008). Procera is a putative DELLA mutant in tomato (*Solanum lycopersicum*): effects on the seed and vegetative plant. *Journal of Experimental Botany* 59(3), 585–93.

Bauchet, G., Munos, S., Sauvage, C., Bonnet, J., Grivet, L. and Causse, M. (2014). Genes involved in floral meristem in tomato exhibit drastically reduced genetic diversity and signature of selection. *BMC Plant Biology* 14, 279.

Beaulieu, J., Doerksen, T. K., MacKay, J., et al. (2014). Genomic selection accuracies within and between environments and small breeding groups in white spruce. *BMC Genomics* 15(1), 1–16.

Bellec-Gauche, et al. (2015). Case Study: multidimensional comparison of local and global fresh tomato supply chains. *GLAMUR Project Report*, p. 56.

Bernacchi, D., Beck-Bunn, T., Emmatty, D., Eshed, Y., Inai, S., Lopez, J., Petiard, V., Sayama, H., Uhlig, J., Zamir, D. and Tanksley, S. (1998b). Advanced backcross QTL analysis of tomato. II. Evaluation of near-isogenic lines carrying single-donor introgressions for desirable wild QTL-alleles derived from *Lycopersicon hirsutum* and *L. pimpinellifolium*. *Theoretical and Applied Genetics* 97, 170–80.

Bernacchi, D., Beck-Bunn, T., Eshed, Y., Lopez, J., Petiard, V., Uhlig, J., Zamir, D. and Tanksley, S. (1998a). Advanced backcross QTL analysis in tomato. I. Identification of QTLs for traits of agronomic importance from *Lycopersicon hirsutum*. *Theoretical and Applied Genetics* 97, 381–97.

Bernardo, R. (2008). Molecular markers and selection for complex traits in plants: learning from the last 20 years. *Crop Science* 48(5), 1649–64.

Bernardo, R. and Yu, J. (2007). Prospects for Genomewide Selection for Quantitative Traits in Maize All rights reserved. *Crop Science* 47(3), 1082–90.

Bertin, N., Guichard, S., Leonardi, C., Longuenesse, J. J., Langlois, D. and Navez, B. (2000). Seasonal evolution of the quality of fresh glasshouse tomatoes under Mediterranean conditions, as affected by air vapour pressure deficit and plant fruit load. *Annals of Botany* 85(6), 741–50.

Bertin, N., Martre, P., Génard, M., Quilot, B. and Salon, C. (2009). Under what circumstances can process-based simulation models link genotype to phenotype for complex traits? Case-study of fruit and grain quality traits. *Journal of Experimental Botany*, 377.

Blanca, J., Canizares, J., Cordero, L., et al. (2012). Variation Revealed by SNP Genotyping and Morphology Provides Insight into the Origin of the Tomato. *PLoS ONE* 7(10), e48198.

Blanca, J., Montero-Pau, J., Sauvage, C., et al. (2015). Genomic variation in tomato, from wild ancestors to contemporary breeding accessions. *BMC Genomics* 16(1), 257.

© Burleigh Dodds Science Publishing Limited, 2017. All rights reserved.

Bolger, A., Scossa, F., Bolger, M. E., et al. (2014). The genome of the stress-tolerant wild tomato species *Solanum pennellii*. *Nature Genetics* 46(9), 1034–8.

Brummell, D. A. and Harpster M.H. (2001). Cell wall metabolism in fruit softening and quality and its manipulation in transgenic plants. *Plant Molecular Biology* 47, 311–39.

Busch, B. L., Schmitz, G., Rossmann, S., Piron, F., Ding, J., Bendahmane, A. and Theres, K. (2011). Shoot branching and leaf dissection in tomato are regulated by homologous gene modules. *The Plant Cell* 23(10), 3595–609.

Cantu, D., Vicente, A. R., Greve L.C., et al. (2008). The intersection between cell wall disassembly, ripening, and fruit susceptibility to Botrytis cinerea. *Proceedings of the National Academy of Sciences of the USA* 105, 859–64.

Causse, M., Buret, M., Robini, K. and Verschave, P. (2003). Inheritance of nutritional and sensory quality traits in fresh market tomato and relation to consumer preferences. *Journal of Food Science* 68, 2342–50.

Causse, M., Chaïb, J., Lecomte, L., et al. (2007). Both additivity and epistasis control the genetic variation for fruit quality traits in tomato. *Theoretical and Applied Genetics* 115, 429–42.

Causse, M., Duffe, P., Gomez, M. C., Buret, M., Damidaux, R., Zamir, D., Gur, A., Chevalier, C., Lemaire-Chamley, M. and Rothan, C. (2004). A genetic map of candidate genes and QTLs involved in tomato fruit size and composition. *Journal of Experimental Botany* 55, 1671–85.

Causse, M., Friguet, C., Coiret, C., et al. (2010). Consumer preferences for fresh tomato at the European scale: a common segmentation on taste and firmness. *Journal of Food Science* 75, S531–41.

Causse, M., Stevens, R., Ben Amor, B., et al. (2011). Breeding for fruit quality. In Jenks, M. and Bebelli P. J. (eds), Breeding for Fruit Quality, Wiley Online, pp. 279–305.

Chaïb, J., Devaux, M.F., Grotte, M., Robini, K., Causse, M., Lahaye, M. and Marty, I. (2007). Physiological relationships among physical, sensory, and morphological attributes of texture in tomato fruits. *Journal of Experimental Botany* 58, 1915–25.

Chakrabarti, M., Zhang, N., Sauvage, C., Munos, S., Blanca, J., Canizares, J., Diez, M. J., Schneider, R., Mazurek, M., McClead, J., Causse, M. and van der Knaap, E. (2013). A cytochrome P450 CYP78A regulates a domestication trait in tomato (*Solanum lycopersicum*). *Proceedings of the National Academy of Sciences of the USA* 110(42), 17125–30.

Chapman, N. H., Bonnet, J., Grivet, L., Lynn, J., Graham, N., Smith, R., Sun, G., Walley, P. G., Poole, M., Causse, M., King, G. J., Baxter, C. and Seymour, G. B. (2012). High-resolution mapping of a fruit firmness-related quantitative trait locus in tomato reveals epistatic interactions associated with a complex combinatorial locus. *Plant Physiology* 159, 1644–57.

Chen, F. Q., Foolad, M. R., Hyman, J., St. Clair, D. A. and Beelman R.B. (1999). Mapping of QTLs for lycopene and other fruit traits in a *Lycopersicon esculentum* 3 *L. pimpinellifolium* cross and comparison of QTLs across tomato species. *Molecular Breeding* 5, 283–99.

Chen, G. P., Hackett, R., Walker, D., et al. (2004). Identification of a specific isoform of tomato lipoxygenase (TomloxC) involved in the generation of fatty acid-derived flavor compounds. *Plant Physiology* 136, 2641–51.

Chen, R.-D. and Tabaeizadeh, Z. (1992b). Expression and molecular cloning of drought-induced genes in the wild tomato *Lycopersicon chilense*. *Biochemistry and Cell Biology* 70, 199–206.

© Burleigh Dodds Science Publishing Limited, 2017. All rights reserved.

Chetelat, R. T., Klann, E., DeVerna, J. W., Yelle, S. and Bennett A.B. (1993). Inheritance and genetic mapping of fruit sucrose accumulation in *Lycopersicon chmielewskii*. *The Plant Journal* 4, 643–50.

Cong B, Barrero, L. S. and Tanksley S.D. (2008). Regulatory change in YABBY-like transcription factor led to evolution of extreme fruit size during tomato domestication. *Nature Genetics* 40, 800–4

Cosgrove D.J. (1998). Cell wall loosening by expansins. *Plant Physiology* 118, 333-9.

Costa, J. M., Ortuño, M. F. and Chaves, M. M. (2007). Deficit irrigation as a strategy to save water: physiology and potential application to horticulture. *Journal of Integrative Plant Biology* 49(10), 1421–34.

Davies, J. N. and Hobson G.E. (1981). The constituents of tomato fruit – The influence of environment, nutrition and genotype. *Critical Review of Food Science and Nutrition* 15, 205–80.

de Vicente, M. C. and Tanksley S.D. (1993). QTL analysis of transgressive segregation in an interspecific tomato cross. *Genetics* 134, 585–96.

Dinar, M. and Stevens M.A. (1981). The relationship between starch accumulation and soluble solids content of tomato fruits. *Journal of the American Society for Horticultural Science* 106, 415–18.

Doganlar, S., Frary, A., Ku, H-K. and Tanksley S.D. (2002). Mapping quantitative trait loci in inbred backcross lines of *Lycopersicon pimpinellifolium* (LA1589). *Genome* 45, 1189-202.

Dorais, M., Papadopoulos, A. P. and Gosselin, A. (2001). Greenhouse tomato fruit quality. *Horticulture Review* 26, 239–319.

Duangjit, J., Causse, M. and Sauvage, C. (2016). Efficiency of genomic selection for tomato fruit quality. *Molecular Breeding* 36(3), 1–16.

El-Soda, M., Malosetti, M., Zwaan, B. J., Koornneef, M. and Aarts, M. G. (2014). Genotype 3 environment interaction QTL mapping in plants: lessons from Arabidopsis. *Trends in Plant Science* 19(6), 390–8.

Eshed, Y. and Zamir, D. (1995). An introgression line population of *Lycopersicon pennellii* in the cultivated tomato enables the identification and fine mapping of yield associated QTL. *Genetics* 141, 1147–62.

Eshed, Y. and Zamir, D. (1996). Less-than-additive epistatic interactions of quantitative trait loci in tomato. *Genetics* 143, 1807–17.

Fereres, E. and Soriano, M. A. (2007). Deficit irrigation for reducing agricultural water use. *Journal of Experimental Botany* 58(2), 147–59.

Fishman, S. and Génard, M. (1998). A biophysical model of fruit growth: simulation of seasonal and diurnal dynamics of mass. *Plant, Cell & Environment* 21(8), 739-52.

Fodor, A., Segura, V., Denis, M., et al. (2014). Genome-Wide Prediction Methods in Highly Diverse and Heterozygous Species: Proof-of-Concept through Simulation in Grapevine. *PLoS ONE* 9(11), e110436.

Frary, A., Doganlar, S., Frampton, A., Fulton, T., Uhlig, J., Yates, H. and Tanksley, S. (2003). Fine mapping of quantitative trait loci for improved fruit characteristics from *Lycopersicon chmielewskii* chromosome 1. *Genome* 46, 235–43.

Frary, A., Fulton, T. M., Zamir, D. and Tanksley S.D. (2004). Advance backcross QTL analysis of a *Lycopersicon esculentum* x L. pennellii cross and identification of possible orthologs in the Solanaceae. *Theoretical and Applied Genetics* 108, 485–96.

Frary, A., Nesbitt, T. C., Frary, A., Grandillo, S., Van der Knaap, E., Cong, B., Liu, J., Meller, J., Elber, R., Alpert, K. B. and Tanksley S.D. (2000). fw-2.2: a quantitative trait locus key to the evolution of tomato fruit size. *Science* 289:85–8

© Burleigh Dodds Science Publishing Limited, 2017. All rights reserved.

Fray, R. G. and Grierson, D. (1993). Identification and genetic analysis of normal and mutant phytoene synthase genes of tomato by sequencing, complementation and co-suppression. *Plant Molecular Biology* 22(4):589-602.

Fridman, E., Carrari, F., Liu, Y. S., Fernie, A. R. and Zamir, D. (2004). Zooming in on a quantitative trait for tomato yield using interspecific introgressions. *Science* 305, 1786-9.

Fridman, E., Pleban, T. and Zamir, D. (2000). A recombination hotspot delimits a wild-species quantitative trait locus for tomato sugar content to 484 bp within an invertase gene. *Proceedings of the National Academy of Sciences of the USA* 97, 4718-23.

Fridman, E. and Zamir, D. (2003). Functional divergence of a syntenic invertase gene family in tomato, potato, and Arabidopsis. *Plant Physiology* 131, 603-9.

Fulton, T. M., Beck-Bunn, T., Emmatty, D., Eshed, Y., Lopez, J., Petiard, V., Uhlig, J., Zamir, D. and Tanksley S.D. (1997). QTL analysis of an advanced backcross of *Lycopersicon peruvianum* to the cultivated tomato and comparisons with QTLs found in other wild species. *Theoretical and Applied Genetics* 95, 881-94.

Fulton, T. M., Bucheli, P., Voirol, E., Lopez, J., Petiard, V. and Tanksley S.D. (2002). Quantitative trait loci (QTL) affecting sugars, organic acids and other biochemical properties possibly contributing to flavor, identified in four advanced backcross populations of tomato. *Euphytica* 127, 163-77.

Fulton, T. M., Grandillo, S., Beck-Bunn, T., Fridman, E., Frampton, A., Lopez, J., Petiard, V., Uhlig, J., Zamir, D. and Tanksley S.D. (2000). Advanced backcross QTL analysis of a *Lycopersicon esculentum x Lycopersicon parviflorum* cross. *Theoretical and Applied Genetics* 100, 1025-42.

Gautier, H., Diakou-Verdin, V., Bénard, C., Reich, M., Buret, M., Bourgaud, F. and Génard, M. (2008). How does tomato quality (sugar, acid, and nutritional quality) vary with ripening stage, temperature, and irradiance?. *Journal of Agricultural and Food Chemistry*, 56(4), 1241-50.

Gautier, H., Lopez-Lauri, F., Massot, C., Murshed, R., Marty, I., Grasselly, D. and Genard, M. (2010). Impact of ripening and salinity on tomato fruit ascorbate content and enzymatic activities related to ascorbate recycling. *Functional Plant Science and Biotechnology*, 4(1), 66-75.

Giovannoni, J. J. (2004). Genetic regulation of fruit development and ripening. *The Plant Cell* 16, 170-80.

Godt, D. E. and Roitsch, T. (1997). Regulation and tissue-specific distribution of mRNAs for three extracellular invertase isoenzymes of tomato suggests an important function in establishing and maintaining sink metabolism. *Plant Physiology* 115, 273-82.

Goldman, I. L., Paran, I. and Zamir, D. (1995). Quantitative trait locus analysis of a recombinant inbred line population derived from a *Lycopersicon esculentum* 3 *L. cheesmanii* cross. *Theoretical and Applied Genetics* 90, 925-32.

Goulet, C., et al. (2012). Role of an esterase in flavor volatile variation within the tomato clade. *Proceedings of the National Academy of Sciences of the USA* 109, 19009-14.

Grandillo, S., Ku, H. M. and Tanksley S.D. (1999). Identifying the loci responsible for natural variation in fruit size and shape in tomato. *Theoretical and Applied Genetics* 99, 978-87.

Grandillo, S. and Tanksley S.D. (1996). QTL analysis of horticultural traits differentiating the cultivated tomato from the closely related species *Lycopersicon pimpinellifolium*. *Theoretical and Applied Genetics* 92, 935-51.

Guichard, S., Bertin, N., Leonardi, C. and Gary, C. (2001). Tomato fruit quality in relation to water and carbon fluxes. *Agronomie* 21(4), 385-92.

© Burleigh Dodds Science Publishing Limited, 2017. All rights reserved.

Guichard, S., Gary, C., Leonardi, C. and Bertin, N. (2005). Analysis of growth and water relations of tomato fruits in relation to air vapor pressure deficit and plant fruit load. *Journal of Plant Growth Regulation* 24(3), 201–13.

Gur, A., Semel, Y., Osorio, S., Friedmann, M., Seekh, S., Ghareeb, B. and Zamir, D. (2011). Yield quantitative trait loci from wild tomato are predominately expressed by the shoot. *Theoretical and Applied Genetics* 122(2), 405–20.

Hall, A. J., Minchin, P. E., Clearwater, M. J. and Génard, M. (2013). A biophysical model of kiwifruit (Actinidia deliciosa) berry development. *Journal of Experimental Botany* 64, 5473–83.

Hamilton, J. P., Sim, S.-C., Stoffel, K., et al. (2012). Single Nucleotide Polymorphism Discovery In Cultivated Tomato Via Sequencing By Synthesis. *Plant Gene* 5(1), 17–29.

Harker, F. R., Redgwell, R. J., Hallett I.C., et al. (1997). Texture of fresh fruit. *Horticultural Reviews* 20, 121–224.

Henderson, C. (1975). Best linear unbiased estimation and prediction under a selection model. *Biometrics* 31(423), 423.

Hileman, L. C., Sundstrom, J. F., Litt, A., et al. (2006). Molecular and phylogenetic analyses of the MADS-Box gene family in tomato. *Molecular Biology and Evolution* 23, 2245–58.

Ho, L. C. (1996). The mechanism of assimilate partitioning and carbohydrate compartmentation in fruit in relation to the quality and yield of tomato. *Journal of Experimental Botany* 47, 1239–43.

Hobson, G. E. and Bedford, L. (1989). The composition of cherry tomatoes and its relation to consumer acceptability. *Journal of Horticulture Science* 64, 321–9.

Hovav, R., Chehanovsky, N., Moy, M., et al. (2007). The identification of a gene (Cwp1), silenced during Solanum evolution, which causes cuticle microfissuring and dehydration when expressed in tomato fruit. *The Plant Journal* 52(4), 627–39.

Isaacson, T., Ronen, G., Zamir, D., et al. (2002). Cloning of tangerine from tomato reveals a carotenoid isomerase essential for the production of beta-carotene and xanthophylls in plants. *The Plant Cell* 14(2), 333–42.

Ito, Y., Kitagawa, M., Ihashi, N., et al.. (2008). DNA-binding specificity, transcriptional activation potential, and the rin mutation effect for the tomato fruit-ripening regulator RIN. *The Plant Journal* 55, 212–23.

Jansen, R. C., Van Ooijen, J. W., Stam, P., Lister, C. and Dean, C. (1995). Genotype-by-environment interaction in genetic mapping of multiple quantitative trait loci. *Theoretical and Applied Genetics*, 91(1), 33–7.

Jin, S., Chen, C. C. S. and Plant A.L. (2000). Regulation byABAof osmoticstress-induced changes in protein synthesis in tomato roots. *Plant, Cell & Environment* 23, 51–60.

Johansson, L., Haglund, A., Berglund, L., et al. (1999). Preference for tomatoes, affected by sensory attributes and information about growth conditions. *Food Quality and Preference* 10, 289–98.

Johansson, L., Haglund, A., Berglund, L., et al. (1999). Preference for tomatoes, affected by sensory attributes and information about growth conditions. *Food Quality and Preference* 10, 289–98.

Johnstone, P. R., Hartz, T. K., LeStrange, M., Nunez, J. J. and Miyao, E. M. (2005). Managing fruit soluble solids with late-season deficit irrigation in drip-irrigated processing tomato production. *Hortscience* 40(6), 1857–61.

Jonas, E. and de Koning, D-J. (2013). Does genomic selection have a future in plant breeding? *Trends in Biotechnology* 31(9), 497–504.

© Burleigh Dodds Science Publishing Limited, 2017. All rights reserved.

Jones R.A. (1986). Breeding for improved post-harvest tomato quality: genetical aspects. *Acta Horticulturae* 190, 77–87.

Kader, A., Morris, L., Stevens, M. and Albrightholton, M. (1978). Composition and flavor quality of fresh market tomatoes as influenced by some post-harvest handling procedures. *Journal of the American Society for Horticultural Science* 103, 6–13.

Klee, H. J. (2010). Improving the flavor of fresh fruits: genomics, biochemistry, and biotechnology. *New Phytologist* 187, 44–56.

Klee, H. J. and Tieman D.M. (2013). Genetic challenges of flavor improvement in tomato. *Trends in Genetics* 29, 257–62.

Korte, A., Vilhjálmsson, B. J., Segura, V., Platt, A., Long, Q. and Nordborg, M. (2012). A mixed-model approach for genome-wide association studies of correlated traits in structured populations. *Nature Genetics* 44(9), 1066–71.

Kumar, S., ChagnÄ, D., Bink, M. C. A. M., et al. (2012). Genomic selection for fruit quality traits in apple (Malus domestica Borkh.). *PLoS ONE* 7(5): e36674. doi:10.1371/journal.pone.0036674.

Labate, J. A., Grandillo, S., Fulton, T., Muños, S., Caicedo, A.L., Peralta, I., Ji, Y., Chetelat, R.T., Scott, J.W., Gonzalo, M.J., Francis, D., Yang, W., van der Knaap, E., Baldo, A.M., Smith-White, B., Mueller, L.A., Prince, J.P., Blanchard, N.E., Storey, D.B., Stevens, M.R., Robbins, M.D., Fen Wang, J., Liedl, B.E., O'Connell, M.A., Stommel, J.R., Aoki, K., Iijima, Y., Slade, A.J., Hurst, S.R., Loeffler, D., Steine, M.N., Vafeados, D., McGuire, C., Freeman, C., Amen, A., Goodstal, J., Facciotti, D., Van Eck, J. and Causse, M. (2007). 1 Tomato. In C. Kole (Ed.), *Genome Mapping and Molecular Breeding in Plants*, Vol. 5, Springer-Verlag, Berlin Heidelberg, 11–135.

Léchaudel, M. and Joas, J. (2006). Quality and maturation of mango fruits of cv. Cogshall in relation to harvest date and carbon supply. *Crop and Pasture Science* 57(4), 419–26.

Lecomte, L., Saliba-Colombani, V., Gautier, A., Gomez-Jimenez, M. C., Duffé, P., Buret, M. and Causse, M. (2004). Fine mapping of QTLs of chromosome 2 affecting the fruit architecture and composition of tomato. *Molecular Breeding* 13, 1–14.

Lecomte, L., Duffé, P., Buret, M., et al. (2004). Marker-assisted introgression of five QTLs controlling fruit quality traits into three tomato lines revealed interactions between QTLs and genetic backgrounds. TAG *Theoretical and Applied Genetics* V109(3), 658–68.

Li, S., Wang, J. and Zhang, L. (2015). Inclusive composite interval mapping of QTL by environment interactions in biparental populations. *PloS ONE*, 10(7), e0132414.

Lin, T., Zhu, G., Zhang, J., et al. (2014). Genomic analyses provide insights into the history of tomato breeding. *Nature Genetics* 46(11), 1220–6.

Lippman, Z. and Tanksley S.D. (2001). Dissecting the genetic pathway to extreme fruit size in tomato using a cross between the small-fruited wild species *Lycopersicon pimpinellifolium* and *L. esculentum* var. giant heirloom. *Genetics* 158, 413–22.

Lippman, Z. B., Cohen, O., Alvarez J.P., et al. (2008). The making of a compound inflorescence in tomato and related nightshades. *PLoS Biol* (11), e288.

Liu, J. P., Van Eck, J., Cong, B. and Tanksley S.D. (2002). A new class of regulatory genes underlying the cause of pear-shaped tomato fruit. *Proceedings of the National Academy of Sciences of the USA* 99, 13302–6

Liu, H. F., Génard, M., Guichard, S. and Bertin, N. (2007). Model-assisted analysis of tomato fruit growth in relation to carbon and water fluxes. *Journal of Experimental Botany*, 58(13), 3567–80.

© Burleigh Dodds Science Publishing Limited, 2017. All rights reserved.

Mageroy, M. H., et al. (2012). A *Solanum lycopersicum* catechol-Omethyltransferase involved in synthesis of the flavor molecule guaiacol. *The Plant Journal* 69, 1043-51

Manning, K., Tör, M., Poole, M., et al. (2006). A naturally occurring epigenetic mutation in a gene encoding an SBP-box transcription factor inhibits tomato fruit ripening. *Nature Genetics* 38(8), 948–52.

Manning, K., Tor, M., Poole, M., Hong, Y., Thompson, A. J., King, G. J., Giovannoni, J. J. and Seymour G.B. (2006). A naturally occurring epigenetic mutation in a gene encoding an SBP-box transcription factor inhibits tomato fruit ripening. *Nature Genetics* 38, 948–52.

Mao, L., Begum, D., Chuang H.W., et al. (2000). JOINTLESS is a MADS-box gene controlling tomato flower abscission zone development. *Nature* 406(6798), 910–13.

Matas, A. J., Gapper, N. E., Chung M.Y., et al. (2009). Biology and genetic engineering of fruit maturation for enhanced quality and shelf-life. *Current Opinion in Biotechnology* 20, 197–203.

Mathieu, S., Cin, V. D., Fei Z.J., et al. 2009. Flavour compounds in tomato fruits: identification of loci and potential pathways affecting volatile composition. *Journal of Experimental Botany* 60, 325–37.

McGlasson, W. B., Last, J. H., Shaw K.J., et al. (1987). Influence of the non-ripening mutants rin and nor on the aroma of tomato fruit. *Hortscience* 22, 632–4.

Melchinger, A. E., Utz, H. F. and Scho¨n C.C. (1998). QTL mapping using different testers and independent population samples in maize reveals low power of QTL detection and large bias in estimates of QTL effects. *Genetics* 149, 383–403.

Meli, V. S., Ghosh, S., Prabha T.N., et al. (2009). Enhancement of fruit shelf life by suppressing N-glycan processing enzymes. *Proceedings of the National Academy of Sciences of the USA* 107, 2413–18.

Meuwissen, T. H. E., Hayes, B. J. and Goddard, M. E. (2001). Prediction of Total Genetic Value Using Genome-Wide Dense Marker Maps. *Genetics* 157(4), 1819–29.

Minic, Z., Jamet, E., San-Clemente, H., et al. (2009). Transcriptomic analysis of Arabidopsis developing stems: a close-up on cell wall genes. *BMC Plant Biology* 9, 17.

Mitchell, J. P., Shennan, C., Grattan, S. R. and May, D. M. (1991). Tomato fruit yields and quality under water deficit and salinity. *Journal of the American Society for Horticultural Science*, 116(2), 215–21.

Molinero-Rosales, N., Jamilena, M., Zurita, S., et al. (1999). FALSIFLORA, the tomato orthologue of FLORICAULA and LEAFY, controls flowering time and floral meristem identity. *The Plant Journal* 20(6), 685–93.

Molinero-Rosales, N., Latorre, A., Jamilena, M., et al. (2004). SINGLE FLOWER TRUSS regulates the transition and maintenance of flowering in tomato. *Planta* 218(3), 427–34.

Monforte, A. J., Asins, M. J. and Carbonell, E. A. (1996). Salt tolerance in *Lycopersicon* species. IV. Efficiency of marker-assisted selection for salt tolerance improvement. *Theoretical and Applied Genetics* 93(5–6), 765–72.

Monforte, A. J., Asins, M. J. and Carbonell, E. A. (1997a). Salt tolerance in *Lycopersicon* species. V. Does genetic variability at quantitative trait loci affect their analysis?. *Theoretical and Applied Genetics* 95(1–2), 284–93.

Monforte, A. J., Asins, M. J. and Carbonell, E. A. (1997b). Salt tolerance in *Lycopersicon* species VI. Genotype-by-salinity interaction in quantitative trait loci detection: constitutive and response QTLs. *Theoretical and Applied Genetics* 95(4), 706–13.

© Burleigh Dodds Science Publishing Limited, 2017. All rights reserved.

Muños, S., Ranc, N., Botton, E., Bérard, A., Rolland, S., Duffé, P., Carretero, Y., Le Paslier, M. C., Delalande, C., Bouzayen, M., Brunel, D. and Causse, M. (2011). Increase in tomato locule number is controlled by two key SNP located near Wuschel. *Plant Physiology* 4, 2244–54.

Muranty, H. I. N., Troggio, M., Sadok, I. S. B., et al. (2015). Accuracy and responses of genomic selection on key traits in apple breeding. *Horticulture Research* 2, 15060.

Mustilli, A. C., Fenzi, F., Ciliento, R., et al. (1999). Phenotype of the tomato high pigment-2 mutant is caused by a mutation in the tomato homolog of DEETIOLATED1. *The Plant Cell* 11, 145-57.

Nuruddin, M. M., Madramootoo, C. A. and Dodds, G. T. (2003). Effects of water stress at different growth stages on greenhouse tomato yield and quality. *Hortscience*, 38(7), 1389–93.

Okabe, Y., Asamizu, E., Saito, T., Matsukura, C., Ariizumi, T., Brès, C., Rothan, C., Mizoguchi, T. and Ezura, H. (2011). Tomato TILLING technology: development of a reverse genetics tool for the efficient isolation of mutants from Micro-Tom mutant libraries. Plant and Cell Physiology 52(11), 1994–2005.

Osorio, S., Ruan, Y. L. and Fernie, A. R. (2014). An update on source-to-sink carbon partitioning in tomato. *Frontiers in Plant Science*, 5, 516.

Page, D., Gouble, B., Valot, B., et al. (2010). Down-regulated protective proteins in tomato correlating with decreased tolerance to low-temperature storage. *Planta* (in press).

Pascale, S. D., Maggio, A., Fogliano, V., Ambrosino, P. and Ritieni, A. (2001). Irrigation with saline water improves carotenoids content and antioxidant activity of tomato. *The Journal of Horticultural Science and Biotechnology* 76(4), 447–53.

Pascual-Banuls, L., Xu, J., Biais, B., et al. (2013). Deciphering genetic diversity and inheritance of tomato fruit weight and composition through a systems biology approach. *Journal of Experimental Botany.* doi: 10.1093/jxb/ert349.

Pascual, L., Desplat, N., Huang, B. E., Desgroux, A., Bruguier, L., Bouchet, J.-P., Le, Q. H., Chauchard, B., Verschave, P. and Causse, M. (2015). Potential of a tomato MAGIC population to decipher the genetic control of quantitative traits and detect causal variants in the resequencing era. *Plant Biotechnology Journal* 13, 565–77.

Pascual, L., Albert, E., Sauvage, C., Duangjit, J., Bouchet, JP., Bitton, F., Desplat, N., Brunel, D., Le Paslier, MC., Ranc, N., Bruguier, L., Chauchard, B., Verschave, P. and Causse, M. (2016). Dissecting quantitative trait variation in the resequencing era: complementarity of bi-parental, multi-parental and association panels. *Plant Science* 242, 120–30.

Paterson, A. H., Damon, S., Hewitt, J. D., Zamir, D., Rabinowitch, H. D., Lincoln, S. E., Lander, E. S. and Tanksley S.D. (1991). Mendelian factors underlying quantitative traits in tomato: comparison across species, generations, and environments. *Genetics* 127, 181–97.

Paterson, A. H., de Verna, J. W., Lanini, B. and Tanksley S.D. (1990). Fine mapping of quantitative trait loci using selected overlapping recombinant chromosomes, in an interspecies cross of tomato. *Genetics* 124, 735–42.

Paterson, A. H., Lander, E. S., Hewitt, J. D., Peterson, S., Lincoln, S. E. and Tanksley S.D. (1988). Resolution of quantitative traits into Mendelian factors by using a complete linkage map of restriction fragment length polymorphisms. *Nature* 335, 721-6.

Pnueli, L., Carmel-Goren, L., Hareven, D., et al. (1998). The SELF-PRUNING gene of tomato regulates vegetative to reproductive switching of sympodial meristems and is the ortholog of CEN and TFL1. *Development* 125, 1979-89.

© Burleigh Dodds Science Publishing Limited, 2017. All rights reserved.

Poiroux-Gonord, F., Bidel, L. P., Fanciullino, A. L., Gautier, H., Lauri-Lopez, F. and Urban, L. (2010). Health benefits of vitamins and secondary metabolites of fruits and vegetables and prospects to increase their concentrations by agronomic approaches. *Journal of Agricultural and Food Chemistry* 58(23), 12065–82.

Poland, J., Endelman, J., Dawson, J., et al. (2012). Genomic Selection in Wheat Breeding using Genotyping-by-Sequencing. *The Plant Genome* 5(3), 103–13.

Powell, A. L., Nguyen, C. V., Hill, T., et al. (2012). Uniform ripening encodes a Golden 2-like transcription factor regulating tomato fruit chloroplast development. *Science*. 336(6089), 1711–15.

Prudent, M., Lecomte, A., Bouchet, JP., Bertin, N., Causse, M. and Génard, M. (2011). Combining ecophysiological modelling and quantitative trait loci analysis to identify key elementary processes underlying tomato fruit sugar concentration. *Journal of Experimental Botany* 62, 907–11.

Prudent, M., Causse, M., Génard, M., Tripodi, P., Grandillo, S. and Bertin, N. (2009). Genetic and ecophysiological analysis of tomato fruit weight and composition - Influence of carbon availability on QTL detection. *Journal of Experimental Botany* 60(3), 923–37.

Quilot, B., Kervella, J., Génard, M. and Lescourret, F. (2005). Analysing the genetic control of peach fruit quality through an ecophysiological model combined with a QTL approach. *Journal of Experimental Botany* 56(422), 3083–92.

Rambla, J. L., Tikunov, Y. M., Monforte, A. J., et al. (2014). The expanded tomato fruit volatile landscape. *Journal of Experimental Botany* 65, 4613–23.

Ranc, N., Munos, S., Santoni, S., et al. (2008). A clarified position for *Solanum lycopersicum* var. *cerasiforme* in the evolutionary history of tomatoes (solanaceae). *BMC Plant Biology* 8, 130.

Reymond, M., Muller, B., Leonardi, A., Charcosset, A. and Tardieu, F. (2003). Combining quantitative trait loci analysis and an ecophysiological model to analyze the genetic variability of the responses of maize leaf growth to temperature and water deficit. *Plant Physiology* 131(2), 664–75.

Ripoll, J., Urban, L. and Bertin, N. (2016b). The potential of the MAGIC TOM parental accessions to explore the genetic variability in tomato acclimation to repeated cycles of water deficit and recovery. *Frontiers in Plant Science*, 6, https://doi.org/10.3389/fpls.2015.01172.

Ripoll, J., Urban, L., Brunel, B. and Bertin, N. (2016a). Water deficit effects on tomato quality depend on fruit developmental stage and genotype. *Journal of Plant Physiology* 190, 26–35.

Ripoll, J., Urban, L., Staudt, M., Lopez-Lauri, F., Bidel, L. P. and Bertin, N. (2014). Water shortage and quality of fleshy fruits–making the most of the unavoidable. *Journal of Experimental Botany* 65(15), 4097–117.

Rodríguez, G. R., Muños, S., Anderson, C., Sim, SC., Michel, A., Causse, M., McSpadden Gardener, B.B., Francis, D. and van der Knaap, E. (2011). Distribution of SUN, OVATE, LC, and FAS Alleles in Tomato Germplasm and their Effect on Fruit Morphology. *Plant Physiology* 156, 275–85.

Ronen, G., Carmel-Goren, L., Zamir, D., et al. (2000). An alternative pathway to β-carotene formation in plant chromoplasts discovered by map-based cloning of Beta and old-gold color mutations in tomato. *Proceedings of the National Academy of Sciences of the USA* 97, 11102–7.

Ronen, G. L., Cohen, M., Zamir, D., et al. (1999). Regulation of carotenoid biosynthesis during tomato fruit development: expression of the gene for lycopene

© Burleigh Dodds Science Publishing Limited, 2017. All rights reserved.

epsilon-cyclase is down-regulated during ripening and is elevated in the mutant Delta. *The Plant Journal* 17, 341-51.

Rosales, M. A., Ruiz, J. M., Hernández, J., Soriano, T., Castilla, N. and Romero, L. (2006). Antioxidant content and ascorbate metabolism in cherry tomato exocarp in relation to temperature and solar radiation. *Journal of the Science of Food and Agriculture* 86(10), 1545-51.

Rose, J., Catalá, C., Gonzalez-Carranza, Z., et al. (2003). Plant cell wall disassembly. In JKC Rose (Ed.), *The Plant Cell Wall*. Oxford, Blackwell, 264-324.

Rose, J. K. C., Bashir, S., Giovannoni, J. J., Jahn, M. M. and Saravanan R.S. (2004). Tackling the plant proteome: practical approaches, hurdles and experimental tools. *The Plant Journal* 39, 715-33.

Rose, J. K. C., Cosgrove, D. J., Albersheim, P., et al. (2000). Detection of Expansin Proteins and Activity during Tomato Fruit Ontogeny. *Plant Physiology* 123, 1583-92.

Ruggieri, V., Francese, G., Sacco, A., et al. (2014). An association mapping approach to identify favourable alleles for tomato fruit quality breeding. *BMC Plant Biology* 14, 337.

Rutkoski, J. E., Heffner, E. L. and Sorrells, M. E. (2011). Genomic selection for durable stem rust resistance in wheat. *Euphytica* 179(1), 161-73.

Sacco, A., Ruggieri, V., Parisi, M., et al. (2015). Exploring a Tomato Landraces Collection for Fruit-Related Traits by the Aid of a High-Throughput Genomic Platform. *PLoS ONE* 10(9), e0137139.

Saïdou, A. A., Thuillet, A. C., Couderc, M., Mariac, C. and Vigouroux, Y. (2014). Association studies including genotype by environment interactions: prospects and limits. *BMC Genetics* 15(1), 1.

Saladie, M., Rose, J. K. C., Cosgrove, D. J., et al. (2006). Characterization of a new xyloglucan endotransglucosylase/hydrolase (XTH) from ripening tomato fruit and implications for the diverse modes of enzymic action. *The Plant Journal* 47, 282-95.

Saliba-Colombani, V., Causse, M., Langlois, D., Philouze, J. and Buret, M. (2001). Genetic analysis of organoleptic quality in fresh market tomato. 1. Mapping QTLs for physical and chemical traits. *Theoretical and Applied Genetics* 102, 259-72.

Sari-Gorla, M., Calinski, T., Kaczmarek, Z. and Krajewski, P. 1997. Detection of QTL3environment interaction in maize by a least squares interval mapping method. *Heredity* 78, 146-57.

Sato, T., Iwatsubo, T., Takahashi, M., et al. (1993). Intercellular localization of acid invertase in tomato fruit and molecular cloning of a cDNA for the enzyme. *Plant Cell Physiology* 34(2), 263-9.

Sauvage, C., Segura, V., Bauchet, G., et al. (2014). Genome-Wide Association in Tomato Reveals 44 Candidate Loci for Fruit Metabolic Traits. *Plant Physiology* 165, 1120-32.

Schauer, N., Semel, Y., Roessner, U., et al. (2006). Comprehensive metabolic profiling and phenotyping of interspecific introgression lines for tomato improvement. *Nature Biotechnology* 24, 447-54.

Schumacher, K., Schmitt, T., Rossberg, M., et al. (1999). The Lateral suppressor (Ls) gene of tomato encodes a new member of the VHIID protein family. *Proceedings of the National Academy of Sciences of the USA* 96(1), 290-5.

Segura, V., Vilhjalmsson, B. J., Platt, A., et al. (2012). An efficient multi-locus mixed-model approach for genome-wide association studies in structured populations. *Nature Genetics* 44(7), 825-30.

© Burleigh Dodds Science Publishing Limited, 2017. All rights reserved.

Semel, Y., Nissenbaum, J., Menda, N., Zinder, M., Krieger, U., Issman, N., Pleban, T., Lippman, Z., Gur, A. and Zamir, D. (2006). Overdominant quantitative trait loci for yield and fitness in tomato. *Proceedings of the National Academy of Sciences of the USA* 103, 12981-6.

Semel, Y., Schauer, N., Roessner, U., Zamir, D. and Fernie, A. R. (2007). Metabolite analysis for the comparison of irrigated and non-irrigated field grown tomato of varying genotype. *Metabolomics* 3(3), 289-95.

Seymour, G. B., Manning, K., Eriksson, E. M., Popovich, A. H. and King G.J. (2002). Genetic identification and genomic organization of factors affecting fruit texture. *Journal of Experimental Botany* 53, 2065-71.

Seymour, G. B., Ostergaard, L., Chapman, N. H., Knapp, S. and Martin, C. (2013). Fruit Development and Ripening. *Annual Review of Plant Biology* 64, 219-41.

Shackel, K. A., Greve, C., Labavitch, J. M., et al. (1991). Cell turgor changes associated with ripening in tomato pericarp tissue. *Plant Physiology* 97, 814-16.

Sim, S.-C., Van Deynze, A., Stoffel, K., et al. (2012b). High-Density SNP Genotyping of Tomato (*Solanum lycopersicum* L.) Reveals Patterns of Genetic Variation Due to Breeding. *PLoS ONE* 7(9), e45520.

Sim, S.-C., Durstewitz, G., Plieske, J. R. et al. (2012a). Development of a Large SNP Genotyping Array and Generation of High-Density Genetic Maps in Tomato. *PLoS ONE* 7(7), e40563.

Simkin A.J., et al. (2004). The tomato carotenoid cleavage dioxygenase 1 genes contribute to the formation of the flavor volatiles b-ionone, pseudoionone, and geranylacetone. *The Plant Journal* 40, 882-92.

Sinesio, F., Cammareri, M., Moneta, E., et al. (2009). Sensory quality of fresh French and Dutch market tomatoes: a preference mapping study with Italian consumers. *Journal of Food Science* 75, S55-67.

Smith, D. L., Abbott, J. A. and Gross K.C. (2002). Down-regulation of tomato beta-galactosidase 4 results in decreased fruit softening. *Plant Physiology* 129, 1755-62.

Speirs, J., Lee, E., Holt, K., et al. (1998). Genetic manipulation of alcohol dehydrogenase levels in ripening tomato fruit affects the balance of some flavor aldehydes and alcohols. *Plant Physiology* 117, 1047-58.

Stevens M.A. (1986). Inheritance of tomato fruit quality components. *Plant Breeding Reviews* 4, 273-311.

Stevens, R., Page, D., Gouble, B., Garchery, C., Zamir, D. and Causse, M. (2008). Tomato fruit ascorbic acid content is linked with monodehydroascorbate reductase activity and tolerance to chilling stress. *Plant Cell and Environment* 31 (8), 1086-96.

Stikic, R., Popovic, S., Srdic, M., Savic, D., Jovanovic, Z., Prokic, L. J. and Zdravkovic, J. (2003). Partial root drying (PRD): a new technique for growing plants that saves water and improves the quality of fruit. Bulg. J. *Plant Physiology* 29(3-4), 164-71.

Tadmor, Y., Fridman, E., Gur, A., Larkov, O., Lastochkin, E., Ravid, U., Zamir, D. and Lewinsohn, E. (2002). Identification of malodorous, a wild species allele affecting tomato aroma that was selected against during domestication. *Journal of Agricultural and Food Chemistry* 50, 2005-9.

Tanksley S.D. (1993). Mapping polygenes. *Annual Review of Genetics* 27, 205-33.

Tanksley S.D. (2004). The genetic, developmental, and molecular bases of fruit size and shape variation in tomato. *The Plant Cell* 16, S181-S189

Tanksley, S. D., Grandillo, S., Fulton, T. M., Zamir, D., Eshed, T., Pétiard, V., Lopez, J. and Beck-Bunn, T. (1996). Advanced backcross QTL analysis in a cross between an elite

© Burleigh Dodds Science Publishing Limited, 2017. All rights reserved.

processing line of tomato and its wild relative *L. pimpinnellifolium*. *Theoretical and Applied Genetics* 92, 213–24.

The Tomato Genome Consortium (2012). The tomato genome sequence provides insights into fleshy fruit evolution. *Nature* 485(7400), 635–41.

Thompson, A. J. and Corlett J.E. (1995). mRNA levels of four tomato (*Lycopersicon esculentum* Mill. L.) genes are related to fluctuating plant and soil water status. *Plant, Cell & Environment* 18, 773–80.

Thompson, A. J., Jackson, A. C., Parker R.A., et al. (2000). Abscisic acid biosynthesis in tomato: regulation of zeaxanthin epoxidase and 9-cis-epoxycarotenoid dioxygenase mRNAs by light/dark cycles, water stress and abscisic acid. *Plant Molecular Biology* 42(6), 833–45.

Tieman, D. M. and Handa, A. K. (1994). Regulation in pectin methylesterase activity modifies tissue integrity and cation levels in ripening tomato (Lycopersicon esculentum Mill.) fruits. *Plant Physiology* 106, 429-36.

Tieman, D., Zeigler, M., Schmelz, E., et al. (2010). Functional analysis of a tomato salicylic acid methyl transferase and its role in synthesis of the flavor volatile methyl salicylate. *The Plant Journal*ournal 62, 113–23.

Tieman, D. M., Zeigler, M., Schmelz, E. A., Taylor, M. G., Bliss, P., Kirst, M. and Klee H.J. (2006). Identification of loci affecting flavour volatile emissions in tomato fruits. *Journal of Experimental Botany* 57, 887-96.

Tieman, D. M., et al. (2006). Aromatic amino acid decarboxylases participate in the synthesis of the flavor and aroma volatiles 2-phenylethanol and 2-phenylacetaldehyde in tomato fruits. *Proceedings of the National Academy of Sciences of the USA* 103, 8287–92.

Tieman, D. M., et al. (2007). Tomato phenylacetaldehyde reductases catalyze the last step in the synthesis of the aroma volatile 2-phenylethanol. *Phytochemistry* 68, 2660-9.

Tikunov, Y., Molthoff, J., de Vos, R. C., Beekwilder, J., van Houwelingen, A., van der Hooft, J. J., Nijenhuis-de Vries, M., Labrie, C. W., Verkerke, W., van de Geest, H., Viquez Zamora, M., Presa, S., Rambla, J.L., Granell, A., Hall, R.D. and Bovy, A. G. (2013). NON-SMOKY GLYCOSYLTRANSFERASE1 Prevents the Release of Smoky Aroma from Tomato Fruit. *The Plant Cell* 25, 8 3067–78.

Tikunov, Y., Lommen, A., de Vos, C. H. R., et al. (2005). A Novel Approach for Nontargeted Data Analysis for Metabolomics. Large-Scale Profiling of Tomato Fruit Volatiles. *Plant Physiology* 139, 1125–37.

Tikunov, Y. M., Molthoff, J., de Vos, R. C. H., et al. (2013). Non-smoky glycosyl transferase1 prevents the release of smoky aroma from tomato fruit. *The Plant Cell* 25(8), 3067–78.

Tinker, N. A., Mather, D. E., Rossnagel, B. G., Kasha, K. J., Kleinhofs, A. and Hayes P.M. (1996). Regions of the genome that affect agronomic performance in two-row barley. *Crop Science* 36, 1053–62.

Tomato-Genome-Consortium (2012). The tomato genome sequence provides insights into fleshy fruit evolution. *Nature* 485, 635–41.

Tucker, G., Price, A. L. and Berger, B. (2014). Improving the power of GWAS and avoiding confounding from population stratification with PC-Select. *Genetics* 197(3), 1045-9.

Uozumi, A., Ikeda, H., Hiraga, M., Kanno, H., Nanzyo, M., Nishiyama, M. and Kanayama, Y. (2012). Tolerance to salt stress and blossom-end rot in an introgression line, IL8-3, of tomato. *Scientia Horticulturae* 138, 1–6.

van der Knaap, E., Lippman, Z. B. and Tanksley S.D. (2002). Extremely elongated tomato fruit controlled by four quantitative trait loci with epistatic interactions. *Theoretical and Applied Genetics* 104 (2-3), 241-7.

© Burleigh Dodds Science Publishing Limited, 2017. All rights reserved.

van der Knaap, E. and Tanksley S.D. (2003). The making of a bell pepper-shaped tomato fruit: identification of loci controlling fruit morphology in Yellow Stuffer tomato. *Theoretical and Applied Genetics* 107, 139-47.

van Eeuwijk, F. A., Bink, M. C., Chenu, K. and Chapman, S. C. (2010). Detection and use of QTL for complex traits in multiple environments. *Current Opinion In Plant Biology* 13(2), 193-205.

Venter, F. (1977). Solar radiation and vitamin C content of tomato fruits. *Acta Horticulturae*, 58, 121-7.

Verbyla, A. P., Cavanagh, C. R. and Verbyla, K. L. (2014). Whole-Genome Analysis of Multienvironment or Multitrait QTL in MAGIC. *G3: Genes| Genomes| Genetics* 4(9), 1569-84.

Vicente, A. R., Saladie, M., Rose, J. K. C., et al. (2007). The linkage between cell wall metabolism and fruit softening: looking to the future. *Journal of the Science of Food and Agriculture* 87, 1435-48.

Vogel, J. T., Tan, B. C., McCarty D.R., et al. (2008). The carotenoid cleavage dioxygenase 1 enzyme has broad substrate specificity, cleaving multiple carotenoids at two different bond positions. *Journal of Biological Chemistry* 283, 11364-73.

Vrebalov, J., Ruezinsky, D., Padmanabhan, V., et al. (2002). A MADS-box gene necessary for fruit ripening at the tomato ripening-inhibitor (rin) locus. *Science* 296(5566), 343-6.

Vrebalov, J., Ruezinsky, D., Padmanabhan, V., White, R., Medrano, D., Drake, R., Schuch, W. and Giovannoni, J. (2002). A MADS-box gene necessary for fruit ripening at the tomato ripening-inhibitor (Rin) locus. *Science* 296, 343-6.

Weller, J. L., Perrotta, G., Schreuder M.E., et al. (2001). Genetic dissection of blue-light sensing in tomato using mutants deficient in cryptochrome 1 and phytochromes A, B1 and B2. *The Plant Journal* 25(4), 427-40.

Whitaker B.D. (2008). Postharvest flavor deployment and degradation in fruits and vegetables. In Bruckner, B. and Grant Willie, S. (Eds), *Fruit Vegetable Flavour*. CRC Press, Cambridge, UK, 103-31.

Wilkinson, J. Q., Lanahan, M. B., Yen H.C., et al. (1995). An ethylene-inducible component of signal transduction encoded by never-ripe. *Science* 270(5243), 1807-9.

Xiao, H., Jiang, N., Schaffner, E., et al. (2008). A retrotransposon-mediated gene duplication underlies morphological variation of tomato fruit. *Science* 319, 1527-30.

Xu, C., Liberatore, K. L., MacAlister, C. A., Huang, Z., Chu, YH., Jiang, K., Brooks, C., Ogawa-Ohnishi, M., Xiong, G., Pauly, M., Van Eck, J., Matsubayashi, Y., van der Knaap, E. and Lippman, Z.B. (2015). A cascade of arabinosyltransferases controls shoot meristem size in tomato. *Nature Genetics* 47, 784-95.

Xu, J., Ranc, N., Munos, S., et al. (2013). Phenotypic diversity and association mapping for fruit quality traits in cultivated tomato and related species. *Theoretical and Applied Genetics* 126(3), 567-81.

Yamamoto, E., Matsunaga, H., Onogi, A., et al. (2016). A simulation-based breeding design that uses whole-genome prediction in tomato. *Scientific Reports* 6, 19454.

Yelle, S., Chetelat, R. T., Dorais, M., DeVerna, J. W. and Bennett A.B. (1991). Sink metabolism in tomato fruit. IV. Genetic and biochemical analysis of sucrose accumulation. *Plant Physiology* 95, 1026-35.

Yelle, S., Hewitt, J. D., Robinson, N. L., et al. (1988). Sink metabolism in tomato fruit .3. Analysis of carbohydrate assimilation in a wild-species. *Plant Physiology* 87, 737-40.

© Burleigh Dodds Science Publishing Limited, 2017. All rights reserved.

Yousef, G. G. and Juvik J.A. (2001). Evaluation of breeding utility of a chromosomal segment from *Lycopersicon chmielewskii* that enhances cultivated tomato soluble solids. *Theoretical and Applied Genetics* 103, 1022–7.

Yu, J., Pressoir, G., Briggs, W. H., et al. (2006). A unified mixed-model method for association mapping that accounts for multiple levels of relatedness. *Nature Genetics* 38(2), 203–8.

Zanor, M. I., Rambla, J.L., Chaïb, J., Steppa, A., Medina, A., Granell, A., Fernie, A. and Causse, M. (2009). Metabolic characterization of loci affecting sensory attributes in tomato allows an assessment of the influence of the levels of primary metabolites and volatile organic contents. *Journal of Experimental Botany* 60, 2139–54.

Zhang, J., Chen, R., Xiao, J., et al. (2007). A single-base deletion mutation in SlIAA9 gene causes tomato (*Solanum lycopersicum*) entire mutant. *Journal of Plant Research* 120(6), 671–8.

Zhang, J., Zhao, J., Liang, Y., et al. (2016). Genome-wide association-mapping for fruit quality traits in tomato. *Euphytica* 207(2), 439–51.

Zhou, S., Wei, S., Boone, B. and Levy, S. (2007). Microarray analysis of genes affected by salt stress in tomato. *African Journal of Environmental Science and Technology* 1(2), 14–26.

© Burleigh Dodds Science Publishing Limited, 2017. All rights reserved.

Chapter 2

Advances and challenges in strawberry genetic improvement

Chris Barbey and Kevin Folta, University of Florida, USA

1 Introduction

The commercial strawberry (*Fragaria × ananassa*) is among a relatively small set of foods that are both nutritious and widely appreciated for flavour. Strawberry production in the United States has an annual value of US$2.4 billion (USDA, 2018), and trends of increased consumption are expected to continue (Rabobank, 2015). Despite its popularity, consumers feel that there is opportunity for improvement, primarily in terms of improved flavours and shelf life. Strawberry production is the foundation of regional economies, and a central crop of many farms. However, while consumption of fruits and vegetables is encouraged and consumers crave strawberry in the diet, the challenges of sustainable production are tremendous and only increasing.

Strawberries are produced from plants that yield fruits with specific seasonal timing. The June-bearing plants begin yielding fruit in spring, and continue through the longest days of summer. Day-neutral strawberries flower regardless of seasonal cues, and short-day varieties have been adopted for winter production in Florida, Mexico, Spain, Turkey, Morocco and other locales of similar climate. The thousands of acres of each modern variety originate from a single maternal plant that is then clonally propagated by runners bearing daughter plants. Plants for transplant are propagated in off-season nurseries in California or Canada and then transported to production areas.

http://dx.doi.org/10.19103/AS.2018.0040.25
© Burleigh Dodds Science Publishing Limited, 2019. All rights reserved.

The strawberry industry runs on two major production systems. Plasticulture is the dominant production method in commercial strawberry production (Fernandez et al., 2001; Poling, 2005). Soil is mounded into long rows under a meter wide, a drip tape is installed to provide fertilizer and water, and then plastic film is stretched over the top. Fumigants are injected under the plastic or via the drip line to suppress nematodes, fungi and weeds. After the fumigants dissipate the soil is nearly sterile, and the new plants are inserted through slits in the plastic. These production plants are established from vegetative, typically cold-treated daughter plants that grow quickly and begin flowering. The annual replacement of plants mitigates the carryover of pests and diseases (Poling, 2005). In some cases commercial strawberries are planted in matted rows, where their runners provide new plants for future seasons. This approach is more appropriate for colder regions with less pest and pathogen pressure (Tabatabaie and Murthy, 2016).

On the production side, regulations around water and fertilizer use, pesticide application, water availability, phase out of soil fumigants and changing labour practices constrain previously mainstream cultural practices. In strawberry, fertilizer, plastic and fuel are the resource inputs most impacting environmental sustainability (Tabatabaie and Murthy, 2016). A number of 'life cycle assessments' have been performed to analyse the major impacts on sustainable production in a variety of production locations, such as Iran (Khoshnevisan et al., 2013), Australia (Gunady et al., 2012), Italy (Girgenti et al., 2014), Spain and the United Kingdom (Williams et al., 2008). The consensus of these reports identified fuel and plastic as primary sources of greenhouse gas emissions, and a central environmental impact in field production. In 2012 it is estimated that 43 000 tons of polyethylene mulch were used in strawberry production in Italy alone (ISPRA, 2013), most of which ends up in landfills (Razza et al., 2012). New practices using biodegradable plastics have been proposed (Girgenti et al., 2014). Strawberry production in greenhouses requires extensive electrical energy input for growth lighting and climate control, which in some cases can be more environmentally adverse than field-based impacts (Khoshnevisan et al., 2013).

The issues of sustainable strawberry production encompass all three pillars of sustainability - economic, social and environmental. U.S. strawberry growers anticipate increasing competitive and regulatory challenges to profitable domestic production. On the other hand, strawberry production is highly taxing on the environment, and the myriad social aspects of labour, fumigant use, water consumption, equity and distribution remain long-standing issues. The way to satisfy all three pillars of sustainability going forward will lie in agricultural innovations that leverage technological advantages in automated production in synergy with new genetic resources.

Strawberry genetic improvement is achieved primarily through traditional breeding. These hybridization methods have provided outstanding varieties for

© Burleigh Dodds Science Publishing Limited, 2019. All rights reserved.

sustainable production. New varieties satisfy basic production demands. Most modern strawberry production varieties produce many large, red, uniformly shaped fruits, and are more resilient to disease. Modern production varieties are, for the most part, genetically optimized to fruit during ideal regional production windows. Fruits from contemporary strawberry cultivars also must survive long-distance shipping and arrive at retail stores without physical damage or substantial natural decay. Agile strawberry genetic improvement is hindered by the complexity of its octoploid genome. Not only is this genome formidable, scientists are still unravelling the exact constituents of the resident subgenomes, a task that began with cytological studies almost a century ago (reviewed in Folta and Davis, 2006). Breeding of new varieties carries the burden of subgenome dominance and linkage drag, two processes that hinder the ability to consolidate genetic gain into advanced breeding selections.

The purpose of this chapter is to highlight some of the newest innovations in production, while explaining the current methods of genetic improvement. Because breeding has been a central barrier, the opportunities in streamlined variety improvement are the main focus of the chapter. Production has already been optimized within current regulations and limits, so discussions in production are more forward thinking and fantastical. Genetic solutions are available now, are being implemented and stand to place a better quality plant in any production system.

2 Threats and solutions to sustainable production

2.1 Competition

The United States is regarded as the second largest strawberry producer in the world; however imports have significantly decreased market prices and depressed domestic industry expansion over the last decade (FAOSTAT, 2016; Zhang et al., 2016). Florida strawberry growers perceive competition from Mexico as the most serious threat to the U.S. domestic strawberry industry, ahead of government regulation, labour shortages and far ahead of pest and disease control (Guan et al., 2015). Fresh imported strawberries to the United States come almost exclusively from Mexico, as imports have increased fourfold since 1997 with the passing of the North American Free Trade Agreement (NAFTA). Mexico has dramatically increased production in recent years, placing significant pressure on domestic growers particularly in the Florida region which relies on high wintertime market prices (Wu and Guan, 2015).

Other complications, including domestic labour shortages, make it increasingly difficult to compete with Mexican producers. The compounding challenges of depressed prices, and high labour cost or low labour availability have led to harvest becoming economically infeasible in certain situations. It is expected that the strawberry industry will become increasingly stressed by

© Burleigh Dodds Science Publishing Limited, 2019. All rights reserved.

Mexican competition, and further antagonized by various production challenges. It is possible that domestic strawberry market share may continue to erode, with some growers shifting to other crops seeking a greater competitive advantage against imports. However, California's strawberry industry maintained its 2016 acreage into 2017, breaking several consecutive years of decline. Per capita consumption of strawberries in the United States has continued to increase over two decades, indicating a growing domestic market coincident with increasing competition, perhaps buffering the perceived impact on American growers.

The impact of potential changes in free trade policies on the domestic strawberry industry are difficult to predict. Driscoll's, the largest American strawberry producer with large operations in both California and Mexico, has signalled that tariffs on Mexican imports would not fundamentally alter the trajectory of the strawberry industry (Charles, 2017). The trends towards Mexican production began in the early 1990s, prior to NAFTA. Overall, favourable labour costs and growing climates in Mexican strawberry agriculture present advantages compared to some aspects of U.S. strawberry production, allowing competition for the fresh domestic market.

Increasing global demand for strawberries, particularly in China, represents a potential avenue for growth for U.S. producers via exportation. Access to Chinese markets has recently been granted after over a decade of negotiations with the USDA. The California Strawberry Commission has indicated confidence in a growing export market to China, where strawberry production halts during the summer months. New domestic varieties that are adapted for Chinese export will be required to maximize U.S. competitiveness in strawberry production. This is similar to the shift towards higher yielding and summer-ripening U.S. varieties, in response to early Mexican competition. Additionally, the domestic U.S. strawberry industry may benefit from establishing cultivars that are distinctive to retail consumers, which may allow competition via differentiation in a market where strawberry cultivars are treated as generic by the consumer (Suh et al., 2017). Sustainable growing practices compatible with the standards of the import country, and rigorous certification processes ensuring pest and pathogen-free imports and exports, are critical to sustainable trade. Overall, the domestic U.S. industry faces significant challenges which will require considerable investment and innovation to overcome.

2.2 Labour, and the advent of mechanical harvesting

Strawberry is almost always harvested by hand. The work is performed by low-wage workers, typically immigrants on temporary agricultural work visas. The process typically involves ungloved workers bending to remove partially-ripened fruits from strawberry plants, placing them directly into plastic 'clamshell' cases, which are then stacked in flats for cooling and shipping. Hand harvested strawberry in the United States is unlikely to be

© Burleigh Dodds Science Publishing Limited, 2019. All rights reserved.

sustained in the future, as improved standards for labourers encourage existing trends towards automation and production outsourcing to countries with a competitive advantage for inexpensive labour.

The loss of human labour will give rise to innovations in mechanical harvesting, which has been projected for over fifty years (Booster and Kirk, 1968; Denisen and Buchele, 1967). Methods of robotic harvesting have been explored for decades, with machines designed to identify the peduncle or fruit with machine vision (Tiezhong and Tianjuan, 2004) and then remove appropriately ripe berries (Hayashi et al., 2010). Examination of several cultivars and mechanical methods suggests that any solution will have to take the variety into consideration (Aliasgarian et al., 2015), which opens the door for breeding specifically for mechanical harvesting. Such varieties may present longer peduncles or thinner canopies, allowing machine vision detection of appropriately ripe fruits.

Reducing labour costs is critical for the competitiveness of the domestic U.S. industry, especially for Florida which is in direct competition with Mexican growers. While it may take considerable time and investment to establish a robust mechanized harvesting solution for strawberry, such a system would neutralize Mexico's labour advantage and allow the domestic U.S. industry to compete effectively in the future (Suh et al., 2017).

2.3 Improved chemical controls

Strawberries are prone to fungal and bacterial disease, as well as assault from insect pests and nematodes. Crop protection strategies are required to limit damage to plants before and during production, and additional strategies are needed at the onset of fruiting. Most of the critical diseases in strawberry are fungal, resulting in lower yields due to infections of the crown or petioles. Other pathogens affect fruit appearance or accelerate post-harvest decay. Strawberry remains as one of the most input-dependent crops, and improved management strategies have been devised. New approaches are pest and pathogen specific, are applied in small amounts and are applied less often. Still, there would be benefit in decreasing inputs in the interest of economic and environmental sustainability.

The emergence of pathogen populations that are resistant to important fungicides in strawberry is becoming commonplace. Resistant grey mould (*Botrytis cinerea*) populations in the United States (Fernández-Ortuño et al., 2012), Germany (Leroch et al., 2013), Spain (Dianez et al., 2002) and elsewhere have recently been reported. Quinone-outside Inhibitor (QoI; strobilurin) fungicides have begun to fail to control *Colletotrichum acutatum* populations in the United States (Forcelini et al., 2016), just as succinate dehydrogenase inhibitor fungicides have similarly failed to control *Botrytis cinerea* (Leroux et al., 2010). The emergence of adapted pathogens have provoked changes

© Burleigh Dodds Science Publishing Limited, 2019. All rights reserved.

in strawberry cultural practices, including more judicious usage of fungicides, simultaneous application of alternative controls and shifts towards alternative but less efficacious fungicides (Leroch et al., 2013; Fernández-Ortuño et al., 2015). These strategies and others will help increase the sustainability of the current suite of chemical fungicides going into the future.

However, it is critical that new fungicides, and new fungicide classes, be developed and deployed. It is important that new fungicides be ecologically sustainable, have adequate target specificity and effectively control damaging pathogens. New and efficacious derivatives of existing fungicides for *Botrytis cinerea* have been created that show greater levels of control on resistant populations (Takagaki et al., 2014). Research into the molecular mechanisms of fungicidal compounds, and identifying differences in target protein structure between pathogen populations, is critical to the development of effective derivatives (Xiong et al., 2015). The integration of -omics tools applied to pathogens is expected to accelerate the rational design of new fungicides with novel targets (Cools and Hammond-Kosack, 2013). It is critical that responsible chemical usage and integrated pest management (IPM) be used in tandem to preserve these important compounds and improve environmental sustainability.

Grey mould and *Rhizopus* rot are the principal pathogens responsible for post-harvest decay. Post-harvest treatment with chitosan, benzothiadiazole and other organic compounds is often performed in strawberry agriculture to elicit strawberry defence genes during storage and shipping (Romanazzi et al., 2013). The elicitors acibenzolar-*S*-methyl and harpin protein, when applied prior to harvest, can also induce resistance in strawberry and suppress spider mites (Tomazeli et al., 2016). Further research in defence elicitors, and their induced resistance genetics, will improve strawberry sustainability and reduce reliance on broad, non-selective chemical controls.

New molecular technologies for selective pest and pathogen control are under development, and hold the promise to substantially affect sustainable pest and pathogen management. Spray-induced gene silencing (SIGS) presents a number of exciting advantages over traditional chemical controls, if it can be developed into an effective commercial solution (Majumdar et al., 2017). SIGs involves the exogenous application of double-stranded DNA (dsDNA) to silence genes in pests and pathogens. These dsDNAs are processed into small RNAs, which interfere with the expression of critical pathogen-derived genes and potentially lead to a loss of virulence or pathogen death (Wang and Jin, 2017). Effective pathogen resistance via transgenic RNAi has been demonstrated for a broad number of agricultural pests and pathogens. These include *Botrytis cinerea* (Wang et al., 2016), *Fusarium graminearum* (Machado et al., 2017), and various arthropods (Bolognesi et al., 2012) and nematodes (Huang et al., 2006).

© Burleigh Dodds Science Publishing Limited, 2019. All rights reserved.

In SIGS, dsDNAs designed against pathogen transcripts are applied to the crop exogenously, rather than transgenically, potentially balancing the need for biotechnological solutions with a consumer preference against genetic engineering, in some markets. Also, SIGS is likely to be durable if deployed correctly. It is hypothesized to be very unlikely for pathogens to evolve effective novel resistance to SIGS, as a variety of sequence targets per pathogen may be simultaneously exploited to maximize the effects (Majumdar et al., 2017). If resistance to a specific small inhibitory RNA were to evolve, it would be simple to design and deploy new dsRNAs targeting against the new mutation. In other words, SIGS represents a relatively durable and updatable strategy for pathogen control once a suitable RNA target for SIGS has been identified. While off-target silencing is measurable in some RNAi feeding studies, RNA-RNA interactions are generally much more predictable and testable than enzyme-chemical interactions (Tan et al., 2016). Thus, RNAi-based strategies for pest and pathogen control is more straightforward relative to pesticidal chemicals, as the potential impact on non-target organisms and the environment can be easily monitored (Bachman et al., 2016). Improved technologies for efficient uptake of exogenous dsDNA, as well as the identification of ideal target transcripts, remains an area of active research both in academia and industry (Chu et al., 2014).

SIGS represents a potential form of pesticidal control that is more specific, sustainable and amenable to rational design (Wang and Jin, 2017). The transgenic event MON87411, approved recently by Canada and the United States represents the first commercial application of RNAi against an insect pathogen, and works via silencing of the vacuolar sorting transcript *DvSnf7* in western corn rootworm (Head et al., 2017). Research into effective application methods and efficacious target transcripts is critical for the implementation of SIGS strategies widely in commercial agriculture.

2.4 Disease-resistant strawberry cultivars

Genetic resistance in cultivated strawberry is highly advantageous, as it would provide passive protection against commercially destructive diseases without additional management requirements or chemical inputs. Specific genetic resistance to most strawberry pathogens has been observed in various *Fragaria* × *ananassa* genotypes (Antanaviciute et al., 2015; Vining et al., 2015), as well as in various wild accessions of diploid *Fragaria* (Calis et al., 2015; Paynter, 2015). However, this resistance has not been extensively consolidated or well-optimized into elite production lines. Much work remains for elucidating the genetic basis for resistance at the molecular level. Marker-assisted selection for disease resistance, involving the selection of genetic elements closely associated with resistance, would provide strawberry breeders the ability

© Burleigh Dodds Science Publishing Limited, 2019. All rights reserved.

to rapidly create disease-resistant cultivars for sustainable production. The development of strongly associated disease resistance loci has been reported for a variety of diseases, such as *Phytophthora fragariae* (Haymes et al., 1997) and *Phytophthora cactorum* (Mangandi et al., 2017). The elimination of methyl bromide use in strawberry production has been cited as a strong motivator for the identification of novel genetic sources of disease resistance (Whitaker et al., 2016). However, significant challenges exist for identifying and selecting disease-resistance genetics in strawberry.

Cultivated strawberry possesses a complex auto-allooctoploid genome, which substantially impedes efforts to identify and reliably consolidate desirable genetics via traditional selection (Folta and Davis, 2006). Additionally, disease resistance is often observed to be closely linked to alleles that are unfavourable for commercial production. Linkage drag is a substantial problem in strawberry breeding, as recombinants that have broken linkage are often still of low utility for breeding as favourable combinations of alleles are seldom to arise in an octoploid context (Stegmeir et al., 2010). Therefore, the ability to screen a very large number of progeny based on genetics, at an early developmental stage, would be highly advantageous for the strawberry breeder. If the causal genetics discriminating disease resistance can be resolved to the allelic level, horizontal transfer tools such as transgenics or gene editing may be used to quickly create production-ready cultivars without the penalties of linkage drag.

Individual genes added to strawberry via transgenesis have been shown to confer resistance disease, including chitinase genes (Asao et al., 1997; Vellicce et al., 2006; Chalavi et al., 2003), thaumatin genes (Schestibratov and Dolgov, 2005), *Npr1* (Silva et al., 2015), β-glucanase genes (Mercado et al., 2015) and the *ELONGATOR* 3 and 3 genes (Silva et al., 2017). Unfortunately, many resistance-associated traits in strawberry are themselves unsuitable for commercialization. For example, increased cell wall thickness is associated with disease resistance due to decreased permeability of pathogenic effectors, but also an undesirable fruit texture (Cantu et al., 2007). Selection for disease resistance in strawberry must be a holistic choice with consideration given to many areas of priority. A refocus on classes of disease-resistance genes which do not typically incur penalties to other traits may assist in this effort.

Several specific genes in strawberry have been associated with disease resistance, such as the Mannose-binding lectin protein associated with *C. acutatum* resistance (Guidarelli et al., 2014). Using new tools in molecular genetics, including targeted sequence capture and QTL analysis in pedigreed populations, several new large-effect QTL resolved to the subgenomic level. These include resistance to angular leaf spot caused by *Xanthomonas fragariae* (*RXf1*), crown and root rot caused by *Phytophthora cactorum* (*Pc1*) and crown rot caused by *Colletotrichum gloeosporioides* (*Cg1*) (Whitaker et al., 2016). These loci provide a high level of resistance, and represent an opportunity

© Burleigh Dodds Science Publishing Limited, 2019. All rights reserved.

to substantially improve genetic resilience to biotic stress. These new methodologies for identifying the genetics of strawberry disease resistance, and other traits, are discussed in-depth in the 'Next steps in genetics' section.

2.5 Emerging pests

A wide variety of animal pests routinely affect strawberry agriculture, including the two-spotted spider mite, spotted wing *Drosophila*, flower thrips, fall armyworm and various secondary pests such as aphids, seed bugs, whiteflies and root weevils. Soil fumigants and foliar insecticides are used extensively in strawberry agriculture to combat these pests. Many pesticidal products have been approved for controlling these various strawberry agricultural pests, and are commonly used proactively and in combination to achieve a broad degree of control. Considerable research effort has been invested in developing sustainable and efficacious pesticides to enhance or replace the existing suite of chemical controls. For example, the elimination of the environmentally toxic methyl bromide has led to the adoption of less damaging alternatives, including metam, dimethyl disulphide, dichloropropene and particularly chloropicrin (Ziedan and Farrag, 2016). These compounds are frequently alternated in use to help prevent the evolution of biological resistance to these important chemical pest controls.

Alternative non-fumigant-based methods have been devised. Steam can be used to sterilize soil, however at high expense (Xu et al., 2017). Biofumigants have been proposed. Amending soil with mustards or arugula or mustard seed meal (Mazzola et al., 2017) also has been shown to have the capacity to suppress disease pressure.

A key area of growth is the development of new and highly selective pesticidal compounds to complement non-chemical strategies for broadly sustainable strawberry production. As described in the previous section, pest control via sprayable small RNAs has been demonstrated to control a number of damaging insect and nematode species. This represents a potentially selective and durable form of agricultural pest control, and is an area of highly active research.

IPM is an increasingly critical component of sustainable pest management in strawberry, and decline in the use of chemical pesticides (Strand, 2008). The use of predatory mites to control spider mites, for example, prevents about four applications of miticide each season for the 15% of Florida strawberry farms using this IPM strategy (Mossler, 2015). Common non-chemical practices include planting strawberries that are certified pest-free, however this does not guarantee that significant pest problems will not develop in the field. The intentional destruction of strawberry fields post-harvest is also commonplace, to remove potential food sources for nematode pests that also could harbour

© Burleigh Dodds Science Publishing Limited, 2019. All rights reserved.

insects. Crop rotations and early field fallowing can reduce pest load, and have been shown specifically to reduce sting and root-knot nematode species (Noling, 1999). Sustainable pest management generally must be continual, prophylactic and adaptive. New strategies for IPM will increase actual yields by mitigating loss to disease, and sustainably decrease the ecological and chemical footprint of strawberry agriculture. New pesticides with increased specificity will help mitigate unintended damage to beneficial insects and the environment (Mossler and Nesheim, 2007). Decreasing yield loss to pests and other biotic stresses indirectly translates into more efficient and sustainable use of agricultural inputs (Pradhan et al., 2015).

Enhancing and optimizing strawberry genetics will promote sustainable agricultural practices through passive resistances to plant pests. The insecticidal Bt crystallized protein is applied in organic pest management of strawberry, and is also used notably as a sustainable transgenic strategy for diverse insect control in various other crops over the past twenty years (Bolda et al., 2003; Carrière et al., 2015). Research into transgenic Bt protein expression in strawberry may lead to cultivars with enhanced passive resistance to pests and potentially a concomitant decrease in pesticide application, as has been noted for other crops according to the U.S. National Academy of Sciences report on genetic engineering in agriculture (National Academies of Sciences, Engineering, and Medicine, 2016). Other transgenic strategies, such as the heterologous expression of trypsin inhibitors, may be used to protect strawberries against a host of pests without the application of exogenous pesticide and significantly advantaging growers and the environment. Native strawberry genes associated with pest resistance have also been identified, and are likely to be gradually introgressed into production cultivars through traditional and marker-assisted breeding selection (Roach et al., 2016). Genetic tolerance to mites has been noted in certain strawberry accessions, as have a host of natural inhibitory and repellent compounds produced under biotic stress conditions (Giménez-Ferrer et al., 1994). Many notable volatile organic compounds produced in the ripe strawberry fruit have antimicrobial and animal repellent properties, including the grape-like methyl anthranilate whose biochemical origins have recently been identified (Pillet et al., 2017).

Sustainable IPM must include all material and strategic sources of pest resistance, including optimized cultural practices, sustainable and selective pesticide application, transgenics, and the introgression and consolidation of favourable native genetics into high-yielding production cultivars.

2.6 Emerging pathogens

Diseases in cultivated strawberry are a major limiting factor for production, and are associated with various challenges in strawberry including high production costs, intensive management requirements, adverse ecological impact,

© Burleigh Dodds Science Publishing Limited, 2019. All rights reserved.

sustainability and impact on human health (Maas, 2004). Various microbial and viral pathogens affect nearly all organs of the strawberry plant, including the fruit, foliar tissue and meristems. Common diseases negatively impacting strawberry production include grey mould caused by *Botrytis cinerea*, various *Colletotrichum* diseases, powdery mildew, angular leaf spot, leaf spot, leaf blight, leaf scorch, charcoal rot and *Phytophthora* crown rot (Peres, 2015). Highly intensive cultural practices are used routinely to mitigate pathogen load, including exclusive planting of certified stock, use of plasticulture, high-dose application of natural and synthetic microbicides, sanitation, replanting and crop rotation (Maas, 2004). Different pathogens may predominate depending on cultivar genetics, local climate and production practices.

Despite rigorous and diverse cultural strategies for pathogen control in strawberry, chemical microbicides remain essential for commercial production (Cordova et al., 2017). In particular, fungicides are key for effective control of strawberry diseases on a commercial scale (Will and Krüger, 1999). The use of synthetic chemical controls, even within the safe label usage, is controversial among broad sections of the potential consumer base (Guthman, 2016). Strawberry is regularly cited by non-governmental environmental organizations as being among the crops 'most contaminated' by trace residue of pesticides (Gatewood, 2017). In another aspect, the severity of negative health consequences caused by chronic and acute fungal mycotoxin contamination in fresh fruit produce, even at low doses, is becoming increasingly recognized (Fernández-Cruz et al., 2010; Smith et al., 2017). Fungal-infected strawberry fruits and jams demonstrate accumulation of toxic fungal metabolites, including Patulin and Ochratoxin A, and aflatoxins (Pensala et al., 1978). Considering the high production challenges posed by strawberry's broad pathogen susceptibility, and the environmental and human health considerations posed by chemical control of pathogens and potentially by the pathogens themselves, more effective control strategies are highly desired. Improved control of pathogens for sustainable strawberry production will arise in three principal areas: improved management strategies, improved chemical controls and improved genetic resistance in production cultivars.

2.7 Improved management strategies

2.7.1 Weather-based crop protection

One of the strongest initiatives towards sustainable strawberry production is the Agroclimate Strawberry Advisory System, based in Florida (Pavan et al., 2011). This web-based portal (www.agroclimate.org) analyses data from weather stations all around Florida, and provides guidance to farmers about the likelihood of disease pressure. The agroclimate algorithms provide a disease model output that informs growers of when conditions are most conducive for

© Burleigh Dodds Science Publishing Limited, 2019. All rights reserved.

disease development. This guidance allows farmers to apply the appropriate fungicide with the right timing, resulting in a 50% decrease in the number of applications (Cordova et al., 2017). This resource has been helpful curbing the use of fungicides typically used to combat *Colletotrichum acutatum* and *Botrytis cinera*, two strawberry pathogens which frequently lead to disease in permissive conditions.

2.7.2 Controlled environments

Strawberry production is especially amenable to production in protected environments, such as greenhouses and artificially lit and heated environments. Such environments may rely on hydroponic production. These production methods are known to decrease the cost of transportation and support local economies with high-value fresh fruit (Mok et al., 2014). Dryland areas are particularly projected to benefit from hydroponic indoor fruit production, especially in terms of water usage and greenhouse gas emissions (de Anda and Shear, 2017). Such shifts to greenhouse production mirror the advances that have already taken place in The Netherlands and Belgium, where restrictions on soil fumigants and herbicides have motivated a shift towards protected cultivation. As recently as 2012, there were over 1600 ha under glass (Lieten, 2013), and that acreage likely stands to increase.

The advantages of protected cultivation are consistent with the ideals of sustainable production. Within a protected environment, strawberries may be grown on benches or in gutters, providing ease of harvest and year-round production that matches grower attractive market windows. Artificial lighting can alter flowering time to match market windows (Yoshida et al., 2012). At the same time these systems present new challenges, as the pests and pathogens irrelevant in the field may become an issue when plants are grown in enclosed space (Elad et al., 1996). Specific cultivars have been shown to perform better in these conditions, especially with respect to disease presentation (Fang et al., 2012). Various strategies have been implemented to combat angular leaf spot in controlled environments (Braun and Hildebrand, 2013). Light also has been used to alter disease pressure. UV-B radiation has been shown to suppress the growth of pathogens, including powdery mildew in strawberry. UV-B radiation in strawberry induced the transcription of a suite of known strawberry defence genes, including strawberry phenylalanine ammonia lyase and chalcone synthase, and improving fruit quality via anthocyanin induction (Kanto et al., 2014). Green LED light has been shown to induce disease resistance generally in a variety of crops including strawberry, suppressing spider mites, improving plant vigour and increasing the accumulation of antimicrobial compounds (Kudo and Yamamoto, 2016). Further research into the interactions of controlled light and defence physiology in strawberry is warranted, and may lead to

© Burleigh Dodds Science Publishing Limited, 2019. All rights reserved.

economically viable solutions for pathogen control for indoor production systems and post-harvest storage. While the industry supporting indoor strawberry cultivation is growing in Europe and the United States, new crop protection strategies, production methods and genetically adapted cultivars will be needed to fit the environmental challenges and opportunities.

3 Post-harvest quality

Sustainability must take into consideration the issues of carbon footprint for transport, as well as post-harvest treatment, storage and waste. Post-harvest quality in strawberry is highly dependent upon fruit firmness – the ability to resist compression and mechanical damage during shipping and retail. One critical post-harvest method involves decreasing the ambient storage temperature to near freezing, quickly after harvest. This is done by a process called forced air cooling or hydrocooling, where berries are placed into a chilled water circulator that may contain sanitizing compounds (Tokarskyy et al., 2015). Many other approaches have been used to sustain fruit quality throughout the supply chain, including treatment with ozone (Pérez et al., 1999), essential oils (Sivakumar and Bautista-Baños, 2014), 1-methylcyclopropene (Ku et al., 1999), chitosan (Hernández-Muñoz et al., 2008, Reddy et al., 2000) and aloe vera (Sogvar et al., 2016). These treatments all have proven effects on slowing pathogen growth and maintaining fruit quality.

Improved metabolic content and flavour may provide a source of natural pest resistance that could also benefit healthful consumption. Volatile organic compounds (VOCs) are a critical component of strawberry flavour, along with sugars and acids (Bood and Zabetakis, 2002). These VOC flavour compounds are frequently shown to contribute to pest and pathogen defence as well as human health (Chambers et al., 2013; Vaughn et al., 1993; Goff and Klee, 2006). While appearance in strawberry has improved notably in larger fruit with more attractive colour, a perceived decline in flavour and aroma has been noted by consumers, breeders and growers for decades (Fletcher, 1917; Folta and Klee, 2016). Generally, the aroma intensity of modern cultivars is perceived as inferior to that of wild strawberries (Ulrich and Olbricht, 2014). Flavour has historically been considered a secondary goal in strawberry breeding, however consumer demands for improved flavour and increasing need for non-fungicidal controls in pathogen management have elevated flavour as a priority in many breeding programmes (Faedi et al., 2000; Whitaker, 2011). As discussed previously, product differentiation through the development of cultivars with distinctive and favourable flavours could help increase the competitiveness of the domestic U.S. industry (Suh et al., 2017).

For these reasons, the introgression of genes encoding enzymes required for the production of desirable VOCs has become a major point of focus for

© Burleigh Dodds Science Publishing Limited, 2019. All rights reserved.

modern cultivar improvement (Vandendriessche et al., 2013). However, the specific introgression of desired flavour compounds in strawberry has been difficult for strawberry breeders. For example, the rare strawberry VOC methyl anthranilate was identified as a desired trait in the 1990s due to a broad range of positive qualities including flavour and antifungal activity (Olbricht et al., 2008). The instability of this trait in breeding populations proved to be a major issue, as unpredictable segregation ratios often delayed breeding efforts using traditional methods (Pillet et al., 2017). Despite considerable effort, progress towards introgressing this VOC into commercial varieties has remained slow, and the genetic mechanisms controlling its biosynthesis have not been elucidated.

Recently, a combination of modern assays in genetics, genomics and biochemistry were applied to octoploid strawberry to discover a necessary gene required for this polygenic trait. Using molecular approaches such as these, genetic markers are being designed to facilitate the introgression of post-harvest traits into advanced selections and commercial cultivars. Success in this area will benefit both consumers and strawberry producers.

4 Next steps in genetics

Associating traits with their causative genetics is crucial for developing more robust and sustainable strawberry cultivated varieties. The deductive process of gene-trait association is challenging in cultivated strawberry due to its peculiarly complex auto-allooctoploid genome. Strawberry's genome has been a substantial challenge for selecting and maintaining desired traits simultaneously in breeding programmes. Unexpected segregation behaviours may emerge due to a confluence of autopolyploid/diploid-like segregation of alleles, while other traits (or elements of a polygenic trait) may segregate with allopolyploid behaviour. Second-generation sequencing technologies resulted in rapid advances for simple crop genomes and model organisms. However, these genomics approaches typically provided insufficient resolution to discriminate octoploid strawberry's numerous allelic variants, and as a result, only situationally allowed the discrimination of genetics related to important traits. In the case of strawberry, it is difficult to achieve the necessary resolution for discovery, as numerous paralogous sequence variants occur both within and between homoeologous chromosomes. As such, successful association studies have often been limited to discrete traits underpinned by simple genetics, such as the deciphering of strawberry γ-decalactone biosynthesis via differential and categorical expression of single transcript. While RNAseq, AFLP and other reductive approaches have demonstrated success in delineating functional genetics in strawberry, the majority of agronomically important traits do not follow simple genetic behaviour and therefore must be studied in a more

© Burleigh Dodds Science Publishing Limited, 2019. All rights reserved.

complete genomic context. Recently developed tools for genetic analysis in strawberry are now allowing these challenges to be addressed efficiently. The technologies discussed below will enable a marked increase in the delineation of gene function relating to marketable traits.

4.1 High-throughput genotyping and QTL analysis

High-density and inexpensive genotyping in octoploid strawberry is highly desired to facilitate association genetics approaches for gene discovery. The Affymetrix IStraw35 Axiom® SNP array, first released in 2015, was designed in collaboration with RosBREED and Affymetrix. The array interrogates over 90 000 subgenomic sequence variations identified within the germplasm of several major international breeding institutions. Subsequently, the IStraw35 array was recently released at lower cost (~$50/sample), and consists of the most useful subset of markers found in the IStraw90 array. These SNP/INDEL arrays can rapidly genotype octoploid germplasm, with informative marker densities typically in the tens of thousands. These genotyping platforms have been successfully used to identify many subgenomic QTL in octoploid strawberry. The demonstrated ability to rapidly and cheaply genotype large populations with high-density subgenomic markers promises to revolutionize gene/trait discovery in strawberry. Importantly, the entire process is relatively accessible to inexperienced users and does not require extremely technical or specific knowledge, nor does it rely on proprietary tools or techniques. A basic association genetics pipeline in strawberry can be described simply: isolated genomic DNA is submitted for IStraw90 or IStraw35 genotyping, and data is returned as a single human-readable text file. This file contains one of four possible calls for each marker (homozygous AA, heterozygous AB, homozygous BB or no-call NC). Basic quality control steps may be performed to remove uninformative markers (e.g. homozygous in all genotypes) and calls with impossible results (e.g. AA progeny from AA/BB parents). Basic genome-wide association studies (GWAS) may be performed using freely available tools to correlate genotypes with user-supplied phenotype scores. Positions of IStraw90 and IStraw35 markers in the diploid physical map are publically available, as are several high-quality octoploid genetic maps. The GWAS software can associate hundreds of phenotypes in hundreds of individuals with tens of thousands of markers in minutes, using only a typical consumer laptop. Pedigree-based QTL association offers a superior degree of association strength than GWAS, and may be executed using software such as FlexQTL, albeit with more demanding computational and technical requirements. Major loci controlling important traits for disease resistance and flavour have been described recently in strawberry using these technologies, and the number of discoveries should increase enormously in coming years. Additionally, identified

© Burleigh Dodds Science Publishing Limited, 2019. All rights reserved.

QTL regions may be directly sequenced via capture-based sequencing and/or third-generation sequencing. This additional resolution may define causal mutations, which would be important for developing perfect genetic markers, determining sequences for genetic engineering or genome editing, and for furthering basic biological understanding.

4.2 Targeted capture sequencing

Targeted capture sequencing represents an important and complementary technique for high-resolution genomic analysis of strawberry. This technique involves the bench-top enrichment of a desired genomic or transcriptomic subset, prior to sequencing. This enrichment is accomplished by the design and application of a panel of single-stranded and tagged probes, complementary to a desired nucleic acid sequence. The synthetic probes hybridize to their complementary sequences, which can then be co-purified with the tagged probe to create an enriched library for sequencing. Desired targets may be sequenced with high depth and coverage using this method, for example wide genomic QTL regions or specific gene and transcript families. For example, enriched R-gene sequencing (Renseq) has demonstrated success in expanding the repertoire of R-genes in wild potato, and has assisted in the isolation and cloning of novel disease-resistant genes. In the era of third-generation sequencing, capture techniques such as these are ready to be applied usefully to complex polyploids such as strawberry. In conjunction with QTL analysis, modern sequencing strategies will further enable rapid and high-resolution gene-trait associations. In some contexts, well-designed sequence capture experiments may allow for novel discovery via bulk segregants analysis alone.

4.3 Third-generation sequencing

Third-generation sequencing, typically involving long reads (>1 kb) of medium or high quality, represents a significant advancement for overcoming the sequence assembly challenges posed by auto-allooctoploid strawberry. The ability to sequence long contiguous sequences allows for relatively facile assembly in comparison to short reads from second-generation sequencing platforms. In the case of auto-allooctoploid strawberry, homeologous alleles may be more easily resolved from highly similar paralogs via comparisons of intergenic sequence contained in the same read. High-quality polyploid reference genomes derived from third-generation sequencing technologies, including that of auto-allooctoploid strawberry, will soon be widely and publicly available. Assembly of other octoploid strawberry genomes and transcriptomes will be made much easier by this octoploid reference. Additionally, in some contexts, less expensive second-generation reads may enjoy emergent

© Burleigh Dodds Science Publishing Limited, 2019. All rights reserved.

feasibility for octoploid resequencing, via read-mapping to the octoploid reference. The application of third-generation sequencing on transcriptomes, or Iso-Seq, can provide direct and conclusive evidence for alternative splice variants, as well as the proportion of these splice variants. High-quality fruit transcriptomes, in combination with dense genetic maps, will also enable Expression QTL (eQTL) studies to determine the genetic basis for differential gene expression, between individuals.

4.4 Expression QTL

eQTL studies present the ability to determine the genetic basis of transcriptomic differences. A successful eQTL thus presents the strawberry breeder with the ability to control the transcription of specific genes via marker-assisted selection. Discoveries made with this approach may be very powerful in strawberry, when the genetic associations are derived from fruit transcriptomes of cultivars of dissimilar quality. Additionally, differentially expressed transcripts *not* associated to any genomic region may represent transcripts that are environmentally and/or developmentally regulated. This is a useful determination both for basic biological understanding and for filtering candidate genes for intentional crop improvement. Transcripts that demonstrate eQTL have an improved likelihood of representing an impactful genetic difference between cultivars, as their observed differential expression is not the result of a stochastic biochemical effect or common environmental response, but instead an intentional result of evolved and potentially selected genetic reregulation. Transcripts which demonstrate eQTL may be regarded as superior targets for study and potential genetic improvement.

5 High-throughput phenotyping

With the advent of rapid genomic and transcriptomic data generation on a grand scale, the bottleneck for gene-trait association increasingly falls to limitations in quantifying plant phenotypes (Araus and Cairns, 2014). High-throughput phenotyping solutions are becoming increasingly necessary, both to improve breeding through traditional selection and for integration with genomics data for discovery. Typical concepts for high-throughput and automated phenotyping involve the use of high-quality cameras and trained AI for image analysis. A proof-of-concept phenotyping device in strawberry was recently developed, involving a 3D camera system coupled with built-for-purpose image analysis software (He et al., 2017). This system reliably determined strawberry achene counts, calyx size, colour, height, length, width and volume. Continued improvement in 3D imaging, AI and foreseeable advances in robotics may soon automate large portions of the gene discovery

© Burleigh Dodds Science Publishing Limited, 2019. All rights reserved.

and crop breeding pipeline. Additionally, high-throughput metabolomics phenotyping has advanced with the advent of sophisticated algorithms for accurately comparing raw GC-MS and LC-MS data (Tikunov et al., 2012; Lommen, 2009). High-quality comparisons of full-scan mass spectrometry data will assist breeders in identifying desirable biochemical phenotypes, and facilitate the association of strawberry metabolomics with genetics.

6 Future trends in research

Sustainable strawberry production must satisfy economic needs while meeting or exceeding social and environmental demands. Recent innovations in production and genetics are poised to meet this challenge, and provide even higher-quality fresh fruit with lower costs to producers and less environmental impact.

Decreasing production costs will come from implementation of precision agricultural innovations, such as automated drones or rovers to scout for pests and pathogens. Upon detection, highly localized treatments of specific crop protection strategies will be deployed, confining pests or diseases to a small part of the plot. Pests and pathogens will be detected and neutralized before symptoms are present, based on powerful sensors that detect threats before they become a problem. These approaches limit the spread of disease and pests, and preclude the need for prophylactic treatments with insecticides or antimicrobials, decreasing production costs for farmers and adverse impacts on the environment.

Automated harvesting is an area of very high potential economic value and is currently the subject of considerable research effort. Labour costs at harvest are substantial and only increasing, threatening economic sustainability. At the same time, the work is repetitive, physical and can negatively affect the health of farm labourers, especially over many seasons. The use of automation can speed harvests, eliminates the threat of labour shortages and high labour costs. Automation also meets the social sustainability need of not requiring people to perform hard physical labour in often inhospitable environments. Automation requires the development of specialized mechanical harvesters, but also varieties amenable to mechanical harvest. The co-evolution of these technologies will eliminate the need for manual labour in strawberry production.

The other major production trends will change the way strawberries are grown in and around urban centres. In The Netherlands, parts of Scandinavia and China, strawberries are grown in extensive greenhouse complexes that permit production throughout the year. These scenarios allow growers to produce fruit during high-value windows. Also, the post-harvest supply chain may be significantly shorter, bringing higher-quality products to consumers at decreased price. Strawberry, a small herbaceous plant that produces a

© Burleigh Dodds Science Publishing Limited, 2019. All rights reserved.

high-value fruit, is well-suited for vertical farms and controlled-environment agriculture. In fact, strawberry is likely the highest-value fruit, relative to total plant size and time required for productive yield. Thus, some fraction of the strawberry agricultural sector may be expected to develop towards controlled-environment 'plant factories' capable of providing high-value fruits within urban settings.

Adjustment of genetics will also bring new traits that should favourably reflect in economic sustainability as well as mitigate environmental impacts. Non-commercial wild strawberries and odd heirloom varieties retain genes long lost during intense selection and breeding, and they serve as a living repository for traits that may be re-introduced to modern varieties. The genes (or genomic regions) underlying the foundations of sensory quality and disease resistance are now known, and it is time to coalesce them into superior genetic backgrounds that also maintain the traditionally coveted traits of yield, fruit size and shipping quality. Modern breeding technologies such as gene editing have been initiated in strawberry. These technologies will be used to introduce functional allelic variants that cannot be moved into elite backgrounds via breeding. Traditional genetic engineering tricks may also be used to improve plants as these technologies become more accepted. Genomics-based tools have spurred the detection of new genetic loci associated with traits of interest. In these cases marker-assisted breeding can allow the rapid introgression of wild or heirloom traits into already improved varieties.

Traits will be controllable by the farmer. The next generation will use signals in the form of light signals or possibly small molecules to hasten flowering, trigger ripening, or help plants protect themselves from abiotic or biotic stress by pre-emptively triggering defence pathways, priming plant responses. These innovations are all alive and well in the laboratory today, and simply await an opportunity to thrive in the field.

Together the next frontiers in strawberry will lead to better fruits that tantalize the consumer's interest. Fruits with new colours, flavours, aromas and textures are alive in the breeder's mind, and in some cases in the greenhouse. These attributes will increase demand, leading to higher prices and a competitive edge for growers compared to traditional germplasm. The interest in environmental sustainability, plus the decreased availability of resources like water and phosphorus, will favour the development of efficient varieties that make the best use of resources provided. Next-generation plants will feature stacks of protective genes that allow them to grow in the presence of disease pressure without exogenous pesticides.

Optimal sustainability will be obtained at the intersection of new production techniques and advanced genetics. It is imperative that both be examined and optimized in parallel, with strong collaborative efforts to ensure that all new solutions are complementary and not being developed in isolation.

© Burleigh Dodds Science Publishing Limited, 2019. All rights reserved.

7 Where to look for further information

For professional growers as well as home gardeners, university extension programmes provide excellent informational resources for regional agriculture. For instance, the University of Florida provides an online repository of extension resources for strawberry, encompassing a variety of topics in strawberry agriculture. Principal topics include current cultural practices, common diseases and prevention, fertilization and nutrition, pest management, post-harvest and handling, common cultivated varieties and strategies for optimal water management. Growers, researchers and consumers may also review articles on DNA technologies used to create local varieties, and information for growing strawberries in the Florida home garden.

For researchers, the Genome Database for Rosaceae (GDR) maintains current genomic and molecular marker data for various strawberry species, including *Fragaria* × *ananassa* and *Fragaria vesca*. Genome assemblies and annotations may be accessed and downloaded via FTP or browsed directly on GDR. A comprehensive list of publically available raw sequencing data in *Fragaria* spp. is also maintained on GDR, and linked for download via the NCBI Short Read Archive. Basic tools for sequence comparison are also provided in-browser. Marker sequences and documentation for the Axiom genome-wide strawberry genotyping array are available on the Affymetrix website. Free and open-sourced tools for genome-wide association of traits, including GAPIT and FlexQTL, are available with documentation.

8 References

Aliasgarian, S., Ghassemzadeh, H. R., Moghaddam, M. and Ghaffari, H. 2015. Mechanical damage of strawberry during harvest and postharvest operations. *Acta Technologica Agriculturae*, 18, 1–5.

Antanaviciute, L., Šurbanovski, N., Harrison, N., McLeary, K. J., Simpson, D. W., Wilson, F., Sargent, D. J. and Harrison, R. J. 2015. Mapping QTL associated with *Verticillium dahliae* resistance in the cultivated strawberry (*Fragaria* × *ananassa*). *Horticulture Research*, 2, 15009.

Araus, J. L. and Cairns, J. E. 2014. Field high-throughput phenotyping: The new crop breeding frontier. *Trends in Plant Science*, 19, 52–61.

Asao, H. G., Nishizawa, Y., Arai, S., Sato, T., Hirai, M., Yoshida, K., Shinmyo, A. and Hibi, T. 1997. Enhanced resistance against a fungal pathogen *Sphaerotheca fumuli* in transgenic strawberry expressing a rice chitinase gene. *Plant Biotechnology*, 14, 145–9.

Bachman, P. M., Huizinga, K. M., Jensen, P. D., Mueller, G., Tan, J., Uffman, J. P. and Levine, S. L. 2016. Ecological risk assessment for DvSnf7 RNA: A plant-incorporated protectant with targeted activity against western corn rootworm. *Regulatory Toxicology and Pharmacology*, 81, 77–88.

© Burleigh Dodds Science Publishing Limited, 2019. All rights reserved.

Bolda, M., Tourte, L., Klonsky, K. and Bervejillo, J. E. 2003. Sample costs to produce organic strawberries: Central coast. *Santa Cruz and Monterey Counties University of California Cooperative Extension*.

Bolognesi, R., Ramaseshadri, P., Anderson, J., Bachman, P., Clinton, W., Flannagan, R., Ilagan, O., Lawrence, C., Levine, S., Moar, W., Mueller, G., Tan, J., Uffman, J., Wiggins, E., Heck, G. and Segers, G. 2012. Characterizing the mechanism of action of double-stranded RNA activity against western corn rootworm (*Diabrotica virgifera virgifera* LeConte). *PLoS ONE*, 7, e47534.

Bood, K. G. and Zabetakis, I. 2002. The biosynthesis of strawberry flavor (II): Biosynthetic and molecular biology studies. *Journal of Food Science*, 67, 2-8.

Booster, D. E. and Kirk, D. E. 1968. State of the art and future outlook for mechanical strawberry harvesting. Corvallis, OR: Agricultural Experiment Station, Oregon State University.

Braun, P. G. and Hildebrand, P. D. 2013. Effect of sugar alcohols, antioxidants and activators of systemically acquired resistance on severity of bacterial angular leaf spot (*Xanthomonas fragariae*) of strawberry in controlled environment conditions. *Canadian Journal of Plant Pathology*, 35, 20-6.

Calis, O., Cekic, C., Soylu, S. and Tör, M. 2015. Identification of new resistance sources from diploid wild strawberry against powdery mildew pathogen. *Pakistan Journal of Agricultural Sciences*, 52, 677-83.

Cantu, D., Vicente, A. R., Greve, L. C., Labavitch, J. M. and Powell, A. L. T. 2007. Genetic determinants of textural modifications in fruits and role of cell wall polysaccharides and defense proteins in the protection against pathogens. *Fresh Produce*, 1, 101-10.

Carrière, Y., Crickmore, N. and Tabashnik, B. E. 2015. Optimizing pyramided transgenic Bt crops for sustainable pest management. *Nature Biotechnology*, 33, 161-8.

Chalavi, V., Tabaeizadeh, Z. and Thibodeau, P. 2003. Enhanced resistance to *Verticillium dahliae* in transgenic strawberry plants expressing a *Lycopersicon chilense* chitinase gene. *Journal of the American Society for Horticultural Science*, 128, 747-53.

Chambers, A., Evans, S. and Folta, K. 2013. Methyl anthranilate and gamma decalactone inhibit strawberry pathogen growth and achene germination. *Journal of Agricultural and Food Chemistry*, 61, 12625-33.

Charles, D. 2017. [Interview]. NPR Radio. https://www.npr.org/sections/thesalt/2017/02/16/515380213/why-ditching-nafta-could-hurt-americas-farmers-more-than-mexicos

Chu, C.-C., Sun, W., Spencer, J. L., Pittendrigh, B. R. and Seufferheld, M. J. 2014. Differential effects of RNAi treatments on field populations of the western corn rootworm. *Pesticide Biochemistry and Physiology*, 110, 1-6.

Cools, H. J. and Hammond-Kosack, K. E. 2013. Exploitation of genomics in fungicide research: Current status and future perspectives. *Molecular Plant Pathology*, 14, 197-210.

Cordova, L. G., Amiri, A. and Peres, N. A. 2017. Effectiveness of fungicide treatments following the Strawberry Advisory System for control of Botrytis fruit rot in Florida. *Crop Protection*, 100, 163-7.

De Anda, J. and Shear, H. 2017. Potential of vertical hydroponic agriculture in Mexico. *Sustainability*, 9, 140.

Denisen, E. L. and Buchele, W. F. 1967. Mechanical harvesting of strawberries. *Proceedings of the American Society for Horticultural Science*, 91, 267-73.

© Burleigh Dodds Science Publishing Limited, 2019. All rights reserved.

Dianez, F., Santos, M., Blanco, R. and Tello, J. C. 2002. Fungicide resistance in *Botrytis cinerea* isolates from strawberry crops in Huelva (southwestern Spain). *Phytoparasitica*, 30, 529-34.

Elad, Y., Malathrakis, N. E. and Dik, A. J. 1996. Biological control of Botrytis-incited diseases and powdery mildews in greenhouse crops. *Crop Protection*, 15, 229-40.

Faedi, W., Mourgues, F. and Rosati, C. 2000. Strawberry breeding and varieties: Situation and perspectives. *Acta Horticulturae*, 567, 51-9.

Fang, X., Phillips, D., Verheyen, G., Li, H., Sivasithamparam, K. and Barbetti, M. J. 2012. Yields and resistance of strawberry cultivars to crown and root diseases in the field, and cultivar responses to pathogens under controlled environment conditions. *Phytopathologia Mediterranea*, 51, 69-84.

FAOSTAT. 2016. http://www.fao.org/faostat/en/#search/strawberries

Fernandez, G. E., Butler, L. M. and Louws, F. J. 2001. Strawberry growth and development in an annual plasticulture system. *HortScience*, 36, 1219-23.

Fernández-Cruz, M. L., Mansilla, M. L. and Tadeo, J. L. 2010. Mycotoxins in fruits and their processed products: Analysis, occurrence and health implications. *Journal of Advanced Research*, 1, 113-22.

Fernández-Ortuño, D., Chen, F. and Schnabel, G. 2012. Resistance to pyraclostrobin and boscalid in *Botrytis cinerea* isolates from strawberry fields in the Carolinas. *Plant disease*, 96, 1198-203.

Fernández-Ortuño, D., Grabke, A., Li, X. and Schnabel, G. 2015. Independent emergence of resistance to seven chemical classes of fungicides in *Botrytis cinerea*. *Phytopathology*, 105, 424-32.

Fletcher, S. 1917. *Strawberry Growing*. New York, NY: Macmillan Co.

Folta, K. M. and Davis, T. M. 2006. Strawberry genes and genomics. *Critical Reviews in Plant Sciences*, 25, 399-415.

Folta, K. M. and Klee, H. J. 2016. Sensory sacrifices when we mass-produce mass produce. *Horticulture Research*, 3. doi:10.1038/hortres.2016.32.

Forcelini, B. B., Seijo, T. E., Amiri, A. and Peres, N. A. 2016. Resistance in strawberry isolates of *Colletotrichum acutatum* from Florida to quinone-outside inhibitor fungicides. *Plant Disease*, 100, 2050-6.

Gatewood, B. 2017. Strawberries remain at top of pesticide list, report says (Online). http://www.cnn.com/2017/03/08/health/dirty-dozen-2017/index.html (accessed).

Giménez-Ferrer, R. M., Erb, W. A., Bishop, B. L. and Scheerens, J. C. 1994. Host-pest relationships between the two-spotted spider mite (Acari: Tetranychidae) and strawberry cultivars with differing levels of resistance. *Journal of Economic Entomology* 87(1), 168-75.

Girgenti, V., Peano, C., Baudino, C. and Tecco, N. 2014. From 'farm to fork' strawberry system: Current realities and potential innovative scenarios from life cycle assessment of non-renewable energy use and green house gas emissions. *Science of the Total Environment*, 473-4, 48-53.

Goff, S. A. and Klee, H. J. 2006. Plant volatile compounds: Sensory cues for health and nutritional value? *Science*, 311, 815-19.

Guan, Z., Wu, F., Roka, F. and Whidden, A. 2015. Agricultural labor and immigration reform. *Choices*, 30(4), 1-9.

Guidarelli, M., Zoli, L., Orlandini, A., Bertolini, P. and Baraldi, E. 2014. The *mannose-binding lectin* gene *FaMBL1* is involved in the resistance of unripe strawberry fruits to *Colletotrichum acutatum*. *Molecular Plant Pathology*, 15, 832-40.

© Burleigh Dodds Science Publishing Limited, 2019. All rights reserved.

Gunady, M. G. A., Biswas, W., Solah, V. A. and James, A. P. 2012. Evaluating the global warming potential of the fresh produce supply chain for strawberries, romaine/cos lettuces (*Lactuca sativa*), and button mushrooms (*Agaricus bisporus*) in Western Australia using life cycle assessment (LCA). *Journal of Cleaner Production*, 28, 81–7.

Guthman, J. 2016. Strawberry growers wavered over methyl iodide, feared public backlash. *California Agriculture*, 70, 124–9.

Hayashi, S., Shigematsu, K., Yamamoto, S., Kobayashi, K., Kohno, Y., Kamata, J. and Kurita, M. 2010. Evaluation of a strawberry-harvesting robot in a field test. *Biosystems Engineering*, 105, 160–71.

Haymes, K. M., Henken, B., Davis, T. M. and Vandeweg, W. E. 1997. Identification of RAPD markers linked to a *Phytophthora fragariae* resistance gene (*Rpf1*) in the cultivated strawberry. *Theoretical and Applied Genetics*, 94, 1097–101.

He, J. Q., Harrison, R. J. and Li, B. 2017. A novel 3D imaging system for strawberry phenotyping. *Plant Methods*, 13, 93.

Head, G. P., Carroll, M. W., Evans, S. P., Rule, D. M., Willse, A. R., Clark, T. L., Storer, N. P., Flannagan, R. D., Samuel, L. W. and Meinke, L. J. 2017. Evaluation of SmartStax and SmartStax PRO maize against western corn rootworm and northern corn rootworm: Efficacy and resistance management. *Pest Management Science*, 73, 1883–99.

Hernández-Muñoz, P., Almenar, E., Valle, V. D., Velez, D. and Gavara, R. 2008. Effect of chitosan coating combined with postharvest calcium treatment on strawberry (*Fragaria* × *ananassa*) quality during refrigerated storage. *Food Chemistry*, 110, 428–35.

Huang, G., Allen, R., Davis, E. L., Baum, T. J. and Hussey, R. S. 2006. Engineering broad root-knot resistance in transgenic plants by RNAi silencing of a conserved and essential root-knot nematode parasitism gene. *Proceedings of the National Academy of Sciences*, 103, 14302–6.

ISPRA (Istituto Superiore per la Protezione e la Ricerca Ambientale). 2013. Italian greenhouse gas inventory, 1990–2011 – National Inventory Report no. 177/2013. Available from: http://www.isprambiente.gov.it/files/pubbli-cazioni/rapporti/Rapporto_177_2013.pdf

Kanto, T., Matsuura, K., Ogawa, T., Yamada, M., Ishiwata, M., Usami, T. and Amemiya, Y. 2014. New UV-B lighting system controls powdery mildew of strawberry. *Acta Horticulturae*, 1049, 655–60.

Khoshnevisan, B., Rafiee, S. and Mousazadeh, H. 2013. Environmental impact assessment of open field and greenhouse strawberry production. *European Journal of Agronomy*, 50, 29–37.

Ku, V. V. V., Wills, R. B. H. and Ben-Yehoshua, S. 1999. 1-Methylcyclopropene can differentially affect the postharvest life of strawberries exposed to ethylene. *HortScience*, 34, 119–20.

Kudo, R. and Yamamoto, K. 2016. Induction of plant disease resistance and other physiological responses by green light illumination. In: Kozai, T., Fujiwara, K. and Runkle, E. S. (Eds), *LED Lighting for Urban Agriculture*. Singapore: Springer Singapore.

Leroch, M., Plesken, C., Weber, R. W. S., Kauff, F., Scalliet, G. and Hahn, M. 2013. Gray mold populations in German strawberry fields are resistant to multiple fungicides and dominated by a novel clade closely related to Botrytis cinerea. *Applied and Environmental Microbiology*, 79, 159–67.

© Burleigh Dodds Science Publishing Limited, 2019. All rights reserved.

Leroux, P., Gredt, M., Leroch, M. and Walker, A.-S. 2010. Exploring mechanisms of resistance to respiratory inhibitors in field strains of *Botrytis cinerea*, the causal agent of gray mold. *Applied and Environmental Microbiology*, 76, 6615-30.

Lieten, P. 2013. Advances in strawberry substrate culture during the last twenty years in the Netherlands and Belgium. *International Journal of Fruit Science*, 13, 84-90.

Lommen, A. 2009. MetAlign: Interface-driven, versatile metabolomics tool for hyphenated full-scan mass spectrometry data preprocessing. *Analytical Chemistry*, 81, 3079-86.

Maas, J. L. 2004. Strawberry disease management. In: Naqvi, S. A. M. H. (Eds), *Diseases of Fruits and Vegetables: Volume II*. Dordrecht, The Netherlands: Springer.

Machado, A. K., Brown, N. A., Urban, M., Kanyuka, K. and Hammond-Kosack, K. 2017. RNAi as an emerging approach to control Fusarium Head Blight disease and mycotoxin contamination in cereals. *Pest Management Science*, 74, 790-9.

Majumdar, R., Rajasekaran, K. and Cary, J. W. 2017. RNA Interference (RNAi) as a potential tool for control of mycotoxin contamination in crop plants: Concepts and considerations. *Frontiers in Plant Science*, 8, 200.

Mangandi, J., Verma, S., Osorio, L., Peres, N. A., Van De Weg, E. and Whitaker, V. M. 2017. Pedigree-based analysis in a multiparental population of octoploid strawberry reveals QTL alleles conferring resistance to *Phytophthora cactorum*. *G3: Genes, Genomes, Genetics*, 7, 1707-19.

Mazzola, M., Agostini, A. and Cohen, M. F. 2017. Incorporation of *Brassica* seed meal soil amendment and wheat cultivation for control of *Macrophomina phaseolina* in strawberry. *European Journal of Plant Pathology*, 149, 57-71.

Mercado, J. A., Barceló, M., Pliego, C., Rey, M., Caballero, J. L., Muñoz-Blanco, J., Ruano-Rosa, D., López-Herrera, C., De Los Santos, B., Romero-Muñoz, F. and Pliego-Alfaro, F. 2015. Expression of the β-1,3-glucanase gene *bgn13.1* from *Trichoderma harzianum* in strawberry increases tolerance to crown rot diseases but interferes with plant growth. *Transgenic Research*, 24, 979-89.

Mok, H.-F., Williamson, V. G., Grove, J. R., Burry, K., Barker, S. F. and Hamilton, A. J. 2014. Strawberry fields forever? Urban agriculture in developed countries: A review. *Agronomy for Sustainable Development*, 34, 21-43.

Mossler, M. 2015. Florida crop/pest management profiles: Strawberry. *University of Florida EDIS*, pp. 1-8.

Mossler, M. and Nesheim, O. 2007. Strawberry pest management strategic plan (PMSP). *University of Florida EDIS*, pp. 1-17.

National Academies of Sciences, Engineering, and Medicine. 2016. *Genetically Engineered Crops: Experiences and Prospects*. National Academies Press.

Noling, J. W. 1999. Nematode management in strawberries. University of Florida Cooperative Extension Service, Institute of Food and Agriculture Sciences, EDIS.

Olbricht, K., Grafe, C., Weiss, K. and Ulrich, D. 2008. Inheritance of aroma compounds in a model population of *Fragaria × ananassa* Duch. *Plant Breeding*, 127, 87-93.

Pavan, W., Fraisse, C. W. and Peres, N. A. 2011. Development of a web-based disease forecasting system for strawberries. *Computers and Electronics in Agriculture*, 75, 169-75.

Paynter, M. 2015. Breeding resistance in strawberry cultivars for *Fusarium oxysporum* f. sp. *fragariae*. MPhil Thesis, School of Agriculture and Food Sciences, The University of Queensland.

Pensala, O., Niskanen, A. and Lindroth, S. 1978. Aflatoxin production in black currant, blueberry and strawberry jams. *Journal of Food Protection*, 41, 344-7.

© Burleigh Dodds Science Publishing Limited, 2019. All rights reserved.

Peres, N. 2015. 2015 Florida plant disease management guide: Strawberry. *University of Florida IFAS EDIS Document* (Online).

Pillet, J., Chambers, A. H., Barbey, C., Bao, Z., Plotto, A., Bai, J., Schwieterman, M., Johnson, T., Harrison, B. and Whitaker, V. M. 2017. Identification of a methyltransferase catalyzing the final step of methyl anthranilate synthesis in cultivated strawberry. *BMC Plant Biology*, 17, 147.

Poling, E. B. 2005. An introductory guide to strawberry plasticulture. Department of Horticultural Science, North Carolina State, 30p.

Pradhan, P., Fischer, G., Van Velthuizen, H., Reusser, D. E. and Kropp, J. P. 2015. Closing yield gaps: How sustainable can we be? *PLoS ONE*, 10, e0129487.

Pérez, A. G., Sanz, C., Ríos, J. J., Olias, R. and Olías, J. M. 1999. Effects of ozone treatment on postharvest strawberry quality. *Journal of Agricultural and Food Chemistry*, 47, 1652-6.

Rabobank. 2015. US berries - increasing consumption is not a silver bullet. https://research.rabobank.com/far/en/sectors/regional-food-agri/us-berries.html

Razza, F., Degli Innocenti, F., Peano, C., Girgenti, V. and Bounous, M. 2012. Improving the efficiency of agricultural and food systems by supply chain: Experience from northern Italy. *Proceedings 8th International Conference on Life Cycle Assessment in the Agri-food Sector, INRA-Agrocampus Rennes*, p. 866.

Reddy, M. V. B., Belkacemi, K., Corcuff, R., Castaigne, F. and Arul, J. 2000. Effect of pre-harvest chitosan sprays on post-harvest infection by Botrytis cinerea and quality of strawberry fruit. *Postharvest Biology and Technology*, 20, 39-51.

Roach, J. A., Verma, S., Peres, N. A., Jamieson, A. R., Van De Weg, W. E., Bink, M. C. A. M., Bassil, N. V., Lee, S. and Whitaker, V. M. 2016. FaRXf1: A locus conferring resistance to angular leaf spot caused by *Xanthomonas fragariae* in octoploid strawberry. *Theoretical and Applied Genetics*, 129, 1191-201.

Romanazzi, G., Feliziani, E., Santini, M. and Landi, L. 2013. Effectiveness of postharvest treatment with chitosan and other resistance inducers in the control of storage decay of strawberry. *Postharvest Biology and Technology*, 75, 24-7.

Schestibratov, K. A. and Dolgov, S. V. 2005. Transgenic strawberry plants expressing a thaumatin II gene demonstrate enhanced resistance to *Botrytis cinerea*. *Scientia Horticulturae*, 106, 177-89.

Silva, K. J. P., Brunings, A., Peres, N. A., Mou, Z. and Folta, K. M. 2015. The Arabidopsis NPR1 gene confers broad-spectrum disease resistance in strawberry. *Transgenic Research*, 24, 693-704.

Silva, K. J. P., Brunings, A. M., Pereira, J. A., Peres, N. A., Folta, K. M. and Mou, Z. 2017. The Arabidopsis *ELP3/ELO3* and *ELP4/ELO1* genes enhance disease resistance in *Fragaria vesca* L. *BMC plant biology*, 17, 230.

Sivakumar, D. and Bautista-Baños, S. 2014. A review on the use of essential oils for postharvest decay control and maintenance of fruit quality during storage. *Crop Protection*, 64, 27-37.

Smith, L. E., Prendergast, A. J., Turner, P. C., Humphrey, J. H. and Stoltzfus, R. J. 2017. Aflatoxin exposure during pregnancy, maternal anemia, and adverse birth outcomes. *The American Journal of Tropical Medicine and Hygiene*, 96, 770-6.

Sogvar, O. B., Saba, M. K. and Emamifar, A. 2016. Aloe vera and ascorbic acid coatings maintain postharvest quality and reduce microbial load of strawberry fruit. *Postharvest Biology and Technology*, 114, 29-35.

© Burleigh Dodds Science Publishing Limited, 2019. All rights reserved.

Stegmeir, T. L., Finn, C. E., Warner, R. M. and Hancock, J. F. 2010. Performance of an elite strawberry population derived from wild germplasm of *Fragaria chiloensis* and *F. virginiana*. *HortScience*, 45, 1140–5.

Strand, L. L. 2008. *Integrated Pest Management for Strawberries*. UCANR Publications.

Suh, D. H., Guan, Z. and Khachatryan, H. 2017. The impact of Mexican competition on the US strawberry industry. *International Food and Agribusiness Management Review*, 20, 591-604.

Tabatabaie, S. M. H. and Murthy, G. S. 2016. Cradle to farm gate life cycle assessment of strawberry production in the United States. *Journal of Cleaner Production*, 127, 548–54.

Takagaki, M., Ozaki, M., Fujimoto, S. and Fukumoto, S. 2014. Development of a novel fungicide, pyribencarb. *Journal of Pesticide Science*, 39, 177–8.

Tan, J., Levine, S. L., Bachman, P. M., Jensen, P. D., Mueller, G. M., Uffman, J. P., Meng, C., Song, Z., Richards, K. B. and Beevers, M. H. 2016. No impact of DvSnf7 RNA on honey bee (*Apis mellifera* L.) adults and larvae in dietary feeding tests. *Environmental Toxicology and Chemistry*, 35, 287–94.

Tiezhong, Z. and Tianjuan, Z. 2004. Strawberry harvesting robot: ?. Segmentation of strawberry image by BP neural network [J]. *Journal of China Agricultural University*, 4, 015.

Tikunov, Y. M., Laptenok, S., Hall, R. D., Bovy, A. and De Vos, R. C. H. 2012. MSClust: A tool for unsupervised mass spectra extraction of chromatography-mass spectrometry ion-wise aligned data. *Metabolomics*, 8, 714–18.

Tokarskyy, O., Schneider, K. R., Berry, A., Sargent, S. A. and Sreedharan, A. 2015. Sanitizer applicability in a laboratory model strawberry hydrocooling system. *Postharvest Biology and Technology*, 101, 103-6.

Tomazeli, V. N., Marchese, J. A., Danner, M. A., Perboni, A. T., Finatto, T. and Crisosto, C. H. 2016. Improved resistance to disease and mites in strawberry, through the use of acibenzolar-S-methyl and harpin to enhance photosynthesis and phenolic metabolism. *Theoretical and Experimental Plant Physiology*, 28, 287–96.

Ulrich, D. and Olbricht, K. 2014. Diversity of metabolite patterns and sensory characters in wild and cultivated strawberries. *Journal of Berry Research*, 4, 11–17.

USDA. 2018. Non-citrus fruits and nuts 2017 summary. National Agricultural Statistics Service. ISSN: 1948-2698.

Vandendriessche, T., Geerts, P., Membrebe, B. N., Keulemans, J., Nicolaï, B. M. and Hertog, M. L. A. T. M. 2013. Journeys through aroma space: A novel approach towards the selection of aroma-enriched strawberry cultivars in breeding programmes. *Plant Breeding*, 132, 217–23.

Vaughn, S. F., Spencer, G. F. and Shasha, B. S. 1993. Volatile compounds from raspberry and strawberry fruit inhibit postharvest decay fungi. *Journal of food Science*, 58, 793-6.

Vellicce, G. R., Ricci, J. C. D., Hernandez, L. and Castagnaro, A. P. 2006. Enhanced resistance to Botrytis cinerea mediated by the transgenic expression of the chitinase gene *ch5B* in strawberry. *Transgenic Research*, 15, 57–68.

Vining, K. J., Davis, T. M., Jamieson, A. R. and Mahoney, L. L. 2015. Germplasm resources for verticillium wilt resistance breeding and genetics in strawberry (*Fragaria*). *Journal of Berry Research*, 5, 183–95.

Wang, M. and Jin, H. 2017. Spray-induced gene silencing: A powerful innovative strategy for crop protection. *Trends in Microbiology*, 25, 4–6.

© Burleigh Dodds Science Publishing Limited, 2019. All rights reserved.

Wang, M., Weiberg, A., Lin, F.-M., Thomma, B. P. H. J., Huang, H.-D. and Jin, H. 2016. Bidirectional cross-kingdom RNAi and fungal uptake of external RNAs confer plant protection. *Nature Plants*, 2, 16151.

Whitaker, V. M. 2011. Applications of molecular markers in strawberry. *Journal of Berry Research*, 1, 115-27.

Whitaker, V. M., Lee, S., Osorio, L. F., Verma, S., Roach, J. A., Mangandi, J., Noh, Y. H., Gezan, S. and Peres, N. 2016. Advances in strawberry breeding at the University of Florida. *Acta Horticulturae*, 1156, 1-6.

Will, F. and Krüger, E. 1999. Fungicide residues in strawberry processing. *Journal of Agricultural and Food Chemistry*, 47, 858-61.

Williams, A., Pell, E., Webb, J., Moorhouse, E. and Audsley, E. 2008. Strawberry and tomato production for the UK compared between the UK and Spain. *Proceedings of the 6th International Conference on LCA in the Agri-food Sector*, pp. 254-62.

Wu, F. and Guan, Z. 2015. Modeling the interactions of strawberry commodity and labor markets in the US and Mexico. *Agricultural & Applied Economics Association and Western Agricultural Economics Association Annual Meeting, San Francisco, CA*. Retrieved from http://ageconsearch. umn. edu/bitstream/205887/2/Modeling. Vol. 20.

Xiong, L., Shen, Y.-Q., Jiang, L.-N., Zhu, X.-L., Yang, W.-C., Huang, W. and Yang, G.-F. 2015. Succinate dehydrogenase: An ideal target for fungicide discovery. In: *Discovery and Synthesis of Crop Protection Products*. American Chemical Society.

Xu, Y., Goodhue, R. E., Chalfant, J. A., Miller, T. and Fennimore, S. A. 2017. Economic viability of steam as an alternative to preplant soil fumigation in California strawberry production. *HortScience*, 52, 401-7.

Yoshida, H., Hikosaka, S., Goto, E., Takasuna, H. and Kudou, T. 2012. Effects of light quality and light period on flowering of everbearing strawberry in a closed plant production system. *Acta Horticulturae*, 956, 107-12.

Zhang, L., Gao, Z. and Vassalos, M. 2016. Sustainable consumer groups and their willingness to pay for tangible and intangible attributes of fresh strawberries. *Selected Paper presented at the Southern Agricultural Economics Association Annual Meeting*.

Ziedan, E.-S. H. and Farrag, E. S. H. 2016. Chemical fumigants as alternatives methyl bromide for soil disinfestation of plant integrated pest management. *International Journal of Agricultural Technology*, 12, 321-8.

© Burleigh Dodds Science Publishing Limited, 2019. All rights reserved.

Chapter 3

Monitoring fruit quality and quantity in mangoes

Kerry Walsh and Zhenglin Wang, Central Queensland University, Australia

1 Introduction

Management of the mango crop requires information on many parameters and processes. In this chapter, focus is directed to methods to assess fruit load (quantity) and quality (in its various forms), both infield and postharvest. Further, a decision support system (DSS) that utilises such information to guide management practices is described. This progression follows the dictum, 'Data is not Information, Information is not Knowledge, Knowledge is not Wisdom'. For example, fruit dry matter (DM) can be measured (data), interpreted in terms of fruit maturation (information) and used in guiding harvest timing (knowledge). Imperatives in this progression are knowledge of the biology of the issue and a robust measurement technique. Thus the first requirement is to define what attributes are useful to measure, the second to measure these attributes accurately and cost effectively and the third is to utilise such information within a management system.

Of course, reliable data are the foundation to a knowledge pyramid. The measurement of fruit quantity and quality can be done manually, but analogue measurement techniques that require human involvement are slow and at times unreliable. The dramatic march in the availability of measurement technologies

http://dx.doi.org/10.19103/AS.2017.0026.14
© Burleigh Dodds Science Publishing Limited, 2018. All rights reserved.

that output a digitised signal and the expansion of telecommunication networks over the last decades offers opportunities in all fields, and mango production is no exception.

In this chapter, specifications for taste and maturity are explored, and then relevant measurement technologies are described.

1.1 A definition of quality

'Beauty is in the eye of the beholder', and so is quality. That is, quality has different meanings from different perspectives. Mango fruit quality, for example, can be seen from the perspective of a grower, a packer, a shipper, a retailer or a consumer, with overlapping sets of attributes of value (Walsh, 2015; Fig. 1). Specifications on quality will also be specific to the fruit maturity/ripening stage.

To the consumer of a 'dessert' mango, fruit quality is judged in terms of visual appearance in general (overall colour, presence of skin defects), size, stage of ripeness (judged on colour and firmness) and taste. Of course, taste involves destructive sampling, and so taste cannot be judged in store, except where taste samplers of fruit from the same consignment are provided. Consumers thus evaluate taste after purchase and will rarely return product to store on a bad taste experience, in contrast to a foreign object experience. However, marketing studies routinely report that a bad eating experience will result in loss of repeat purchases by the consumer for approximately six weeks (Diehl et al., 2013). Of course, there are also different consumers of mango, with different requirements. For example, some consumers prefer a redder external colour (e.g. the Chinese New Year market), while others prefer higher volatile levels (a stronger 'mango' aroma). There are also consumers purchasing mango for cooking purposes, for juice and for other purposes. Obviously a juice producer gives little attention to external appearance, except as it reflects internal problems.

The marketer/retailer is responsible for presenting the consumer with aesthetically pleasing dessert fruit that are free of significant defects and that will be ripe within a few days. Thus size, external colour, defects and shelf life are the primary quality attributes at this stage.

The packhouse is a point in the supply chain where sorting at an individual fruit level can occur, with technology available for grading fruit in-line. Preventative treatments can also be applied to fruit on the packline, for example, bactericidal and fungicidal treatments to prevent disorders further along the value chain. Of course, it is also possible to introduce 'downstream' disorders. For example, fruit harvested during rain events have turgid skin cells that can easily be damaged by brushes, etc. in the packline. Likewise, if the wash treatments are set at too high a temperature, skin scalding may result. This damage may not be noticeable in the packhouse

© Burleigh Dodds Science Publishing Limited, 2018. All rights reserved.

but will become apparent some days later, after the fruit has been moved to market. The packer lives in compromise between the specifications of fruit taken from the field, the requirements of the marketer/retailer and consumer.

The grower is challenged to produce fruit that meets the concerns of the packer, the marketer/retailer and the consumer. Out-of-specification fruit can be rejected at the packhouse or indeed later in the value chain, but any rejection is lost profit. The primary tools available to the grower are choice of variety and agronomics, in particular disease control and timing of harvest.

1.2 Defining maturity and ripeness

It is important to have a common language for the sake of effective communication. Unfortunately, the terms 'maturity' and 'ripeness' are often used too loosely. There are four terms of importance - *physiological maturity*, *harvest maturity*, *commercial maturity* and *ripeness*.

Mangoes are climacteric fruit, that is, after *physiological maturity* the fruit are able to undergo a *ripening* process triggered by ethylene, in which respiration rate rises, skin and flesh colours change, starch converts to soluble sugars, organic acid levels decrease, tissue softens and volatiles increase (Fig. 2). As an analogy, reproductive maturity in humans is reached (in early teen years) earlier than it is generally recommended to reproduce. Fruit physiological maturity occurs earlier than normal harvest or *harvest maturity*. Fruits reach *commercial maturity* when the fruit is at a stage acceptable to the consumer - for mango, when the fruit are partly ripened.

The decision on timing of harvest at some time after physiological maturity and before commercial maturity is a commercial decision involving compromise. As long as the fruit remains on the tree, there is opportunity to import photosynthate and increase reserves, thus increasing eating quality after ripening. However, the longer the fruit remains on the tree past physiological maturity, the shorter is its ripening time, and thus, the shorter the postharvest life. For some markets, it may be desired to have fruit ripen on tree, to have maximum carbohydrate accumulation (and taste) but no shelf life. The other extreme is a maximal shelf life and thus transport capability, at the expense of carbohydrate content and eating quality.

Ripening is a continuous process of senescence. The extent of ripening to achieve '*commercial maturity*' will vary by market, so generally fruit are sold in a partly ripened state, allowing the consumer some control over the extent of ripening at consumption. Generally, dessert fruit are consumed at a stage defined by flesh firmness, when starch has not quite fully converted to soluble sugar (Fig. 2).

© Burleigh Dodds Science Publishing Limited, 2018. All rights reserved.

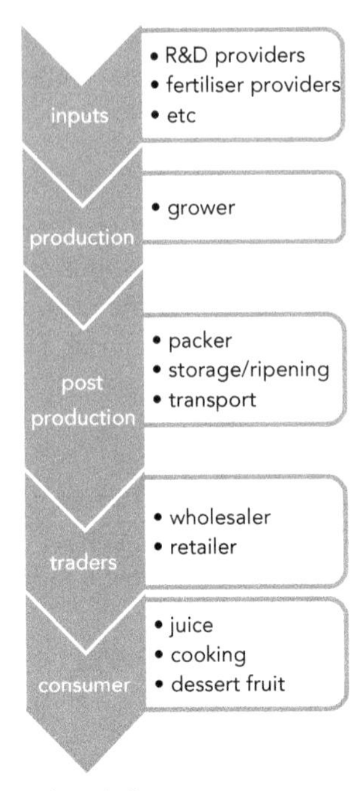

Figure 1 Steps in the mango value chain.

1.3 A taste specification

In general, fruit taste is indexed by soluble sugar content (SSC), organic acid content (TA) and flesh texture. In some fruit, with a low organic acid content and consistent texture (e.g. melon), SSC is the defining attribute. In other fruit, with low sugar content (e.g. lime), the organic acid content is of primary concern. For some fruit, for example, oranges, the soluble sugar:acid ratio is of importance. However, it is not only the ratio of sugars to acid that is relevant, but also the total level (creating a 'mouth feel'). Indeed, Florida oranges were graded to a matrix of SSC:TA ratios defined by the SSC level (Obenland et al., 2009). To address this complexity, Jordan et al. (2001) introduced the BrimA measure, calculated as Brix − k × total acid. This relationship was subsequently adapted slightly in the California standard for citrus, as [SSC – (4*TA)*16.5] (https://www.cacitrusmutual.com/marketing/the-california-standard/ Accessed date: 25 February 2017).

The taste of a mango is defined by its soluble sugar and organic acid content, flesh texture and the profile of volatile compounds. The levels of these attributes are determined by variety, growing and ripening conditions, fruit maturity and ripeness. In mango, SSC rises as starch and organic acid content and firmness decline during ripening. In optimally ripened fruit, fruit are of

© Burleigh Dodds Science Publishing Limited, 2018. All rights reserved.

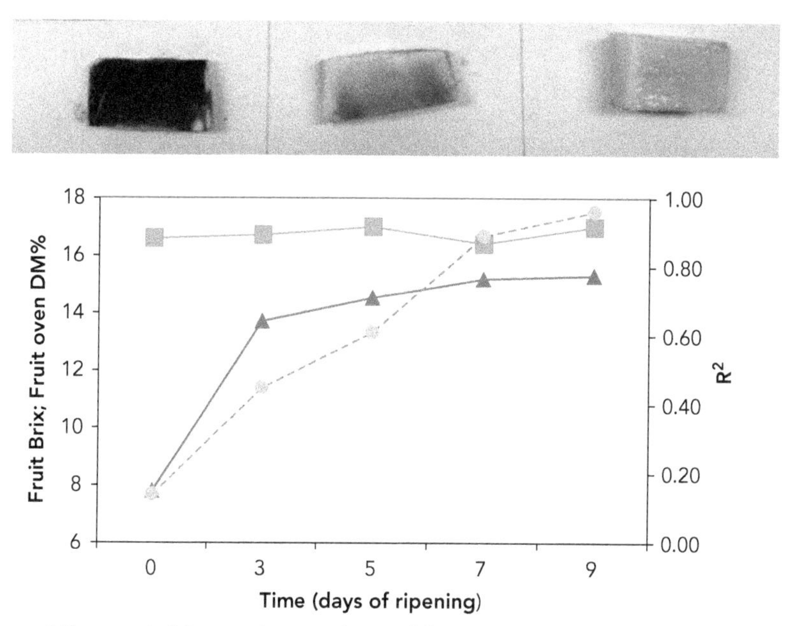

Figure 2 Top panel: IKI stained mango fruit at different stages of ripeness (left, hard green fruit, middle, fruit at eating stage; right, over ripened fruit). Bottom: change in DM (blue square), Brix (red triangle) and and R^2 of Brix-DM correlation (green circles) with time ripened (left axis). *Source* CQUniversity.

optimal firmness and low acid content, such that SSC becomes a dominant determinant of eating quality. Thus SSC of fully ripe fruit can be used as an index for determining eating quality (taste) (Fig. 3a). However, as noted earlier, mangoes are rarely eaten at a completely fully ripe (senescent) stage; rather, they are eaten slightly before full ripening, when they have slightly higher firmness and acidity. Thus the BrimA index also has relevance to mango (Fig. 4).

However, at the time of consumption, there is a higher correlation of eating quality score to DM content than to SSC (Henroid et al., 2014; Fig. 3b). DM thus provides an index of the potential eating quality rather than the stage of ripeness, being a measure of both tissue-soluble and -insoluble solid content. DM is insensitive to a change between starch, organic acid and sugar pools, as occurs during ripening. Soluble solids will include sugars and organic acids. Insoluble solids will include starch, and also all cell structural material (e.g. cell membranes and wall). But as cell structural material remains fairly constant, change in mango tissue DM represents change in soluble sugars and starch levels, with a small contribution from organic acid levels. The DM content of mango tissue is thus well correlated with its SSC when fully ripe (Fig. 6). Therefore DM is a useful index for the determination of eating quality of a mango fruit.

© Burleigh Dodds Science Publishing Limited, 2018. All rights reserved.

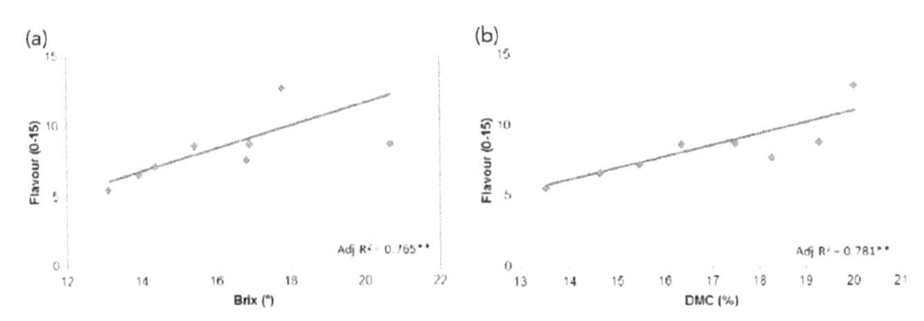

Figure 3 Plot of eating quality score as a function of (a) SSC (Brix) and (b) Dry matter content (DMC). *Source* Campbell (2015).

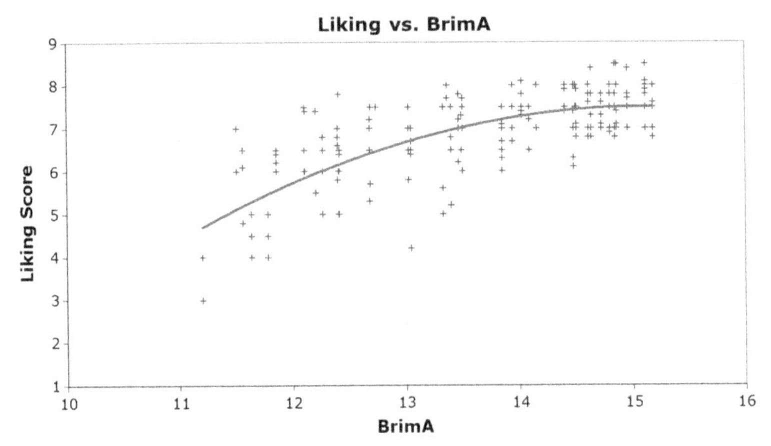

Figure 4 Plot of mango eating quality (scale of 1, extreme dislike, to 9, extreme liking) score vs Brim A.

Source Mark Loeffen, Dyletics P/L.

Indeed, DM at any stage post physiological maturity should be proportional to SSC of fully ripened fruit. Specifically, DM at fruit harvest is well correlated with ripened fruit SSC and eating quality (Whiley et al., 2006; Fig. 5). A caveat applies if the total concentration of fruit starch and sugars changes during ripening - for example, fruit can lose water if ripened under low humidity conditions, leading to an increase in DM. In practice, industry is focused on retaining weight of fruit, which is achieved by maintaining high humidity. Conversely, fruit respiration will result in loss of carbohydrate and thus a tendency for lower DM. For example, if a respiration rate of 7 mL CO_2/kg/h (equivalent to 0.29 mmol/kg/h or 13 mg/kg/h) (Mitcham and McDonald, 1993) were maintained throughout a week of ripening (an overestimate as respiration rate decreases post climacteric), the fruit would lose 0.3 g/kg/day or 2.1 g/kg over seven days. This equates to a DM loss of 0.21%. Thus fruit

© Burleigh Dodds Science Publishing Limited, 2018. All rights reserved.

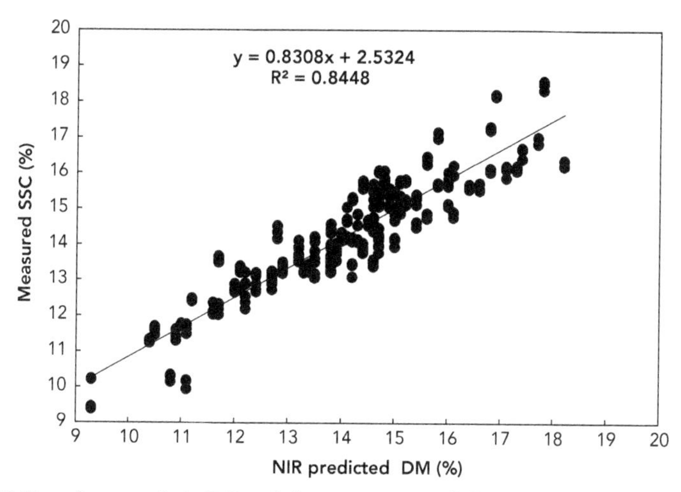

Figure 5 Plot of mango fruit SSC at fully ripe vs DM at fully ripe stage. SSC at fully ripe stage is approximately 1 unit lower than %DM. *Source* CQUniversity.

loss due to respiration during fruit ripening may be measurable but is not a large amount, and likely offset by a small transpirational water loss.

On the basis of taste panel tests (Henroid et al., 2014), the Australian Mango Industry Association has recommended minimum DM levels by variety [15% DM for cultivars B74, Kensington Pride and Honey Gold, and 14% DM for R2E2; AMIA, 2016; http://www.industry.mangoes.net.au/my-mango/ Accessed date: 26 February 17] for ripened fruit to achieve an acceptable eating quality. The US Mango Importers Association has also commissioned taste tests through UC Davis, California, with similar recommendations (C. Crisosto, pers. comm., oral presentation at IHC2014, Brisbane, Australia). These specifications are for a DM value associated with an acceptable eating experience, not a good eating experience, which is associated with higher DM values. Applying this cut-off as a specification for the average of a population is, of course, generous, as 50% of any population is below average.

1.4 A harvest maturity specification

The decision to pick (harvest maturity) (Subedi et al., 2007; Walsh, 2015) is based on an assessment of the following:

- timing - calendar days from flowering
- timing - heat sums from flowering
- fruit shape (broadening of shoulders)
- skin colour (reddening or yellowing)
- internal flesh colour (white to yellow)

© Burleigh Dodds Science Publishing Limited, 2018. All rights reserved.

- DM

However, a number of varieties are well coloured and full in shape well before harvest maturity. Assessment of attributes such as flesh colour requires destructive sampling.

Several indices can be used together to improve decision-making. For example, the DM specification required for eating quality (e.g. 15% DM) can be used as a specification for harvest. This use of DM is separate from its use as an index for eating quality, although obviously the harvest maturity DM specification must have as a minimum the value of the eating quality DM specification. However, the level of DM in a fruit depends on photosynthetic conditions as well as the maturity of the fruit. High photosynthetic rates will favour higher DM levels, as will manipulations such as girdling. Thus the DM level associated with the desired harvest maturity as established by flesh colour, fruit shape or other attributes should be established for a given variety/growing condition, in addition to the requirement to achieve the DM level required for eating quality.

1.5 A ripening specification

Ledger et al. (2012) have provided a useful treatise on mango ripening. In summary, fruit are typically harvested and transported in a hard green condition. Best practice for ripening involves use of a constant temperature, around 20°C. This may occur in transit (in container) for long-distance markets or in ripening centres close to markets. Ripening may be initiated using ethylene, although this may not be necessary in later season fruit or in fruit left longer on tree. Fruit are ripened until a desired firmness level is reached, typically gauged by hand feel and external colour, and then fruit are transported to stores in a condition with some days of shelf life remaining.

2 Monitoring harvest maturity: making the decision to pick

Relying on human expertise to judge harvest maturity is fraught, especially given the typically frantic pace of activity at harvest time. Further, assessing these attributes requires a level of skill that is often absent in a harvest crew. Quantifiable attributes that can be rapidly assessed are required, allowing sampling of a statistically significant number of fruit within an orchard.

2.1 Monitoring flowering

In a 'best practice' orchard, the date and extent of flowering are recorded to inform a heat unit-based model of time to harvest maturity and an estimate of the potential size of the crop. A record of early flowering trees could be used

© Burleigh Dodds Science Publishing Limited, 2018. All rights reserved.

to enable variable rate spraying, with sprays turned off for non-flowering trees, and also to inform selective harvest, with earlier flowering trees having fruit that will mature earlier.

Different systems exist for estimating the date of flowering, from the swollen bud stage or later stages, for example, the 'Christmas tree stage', with two-thirds of flowers open on the inflorescence (Fig. 6a). Typically, assessing the extent of flowering involves a manual assessment of the percentage of terminals that have become reproductive. Flowering assessment can take many forms in commercial practice, for example, a drive through several rows of a block, with a mental estimation of the extent of flowering.

Attempts to introduce machine vision for the estimation of flowering have been made (Wang et al., 2016; Fig. 6a,b). While the colours of inflorescences are reasonably distinct from foliage, enabling segmentation of the image (Fig. 6a), the fact that the number of pixels associated with inflorescences increases due to increase in both the number of flowering terminals and the size of the inflorescences is a difficulty. This is not an issue if there is only a single flowering event; however, if flowering events overlap, identification of the individual events becomes difficult (Fig. 6c). Further progress with multiple flowering events will require the identification of flowering terminals, rather than simple counting of flower-associated pixel number.

Of course there is a large and variable flower and young fruit drop in mango, with only 8% or less of inflorescences setting a marketable fruit. Poor pollination and rain during flowering, with consequent disease issues, are associated with poor fruit set. Thus, in addition to flowering estimation, fruit set estimation is required (see Section 3.1).

2.2 Monitoring time and field temperature

Calendar time from flowering is a base guide to maturation time, based on the experience in previous years. However, development is temperature dependent, so an estimate of maturation time based on heat units will be more accurate. A heat unit is typically based on the summation of the difference between the average of daily maximum and minimum temperatures and a base temperature, for example, 12°C, from the date of flowering. The target value is set by variety and local experience, for example, 1600 heat units is required for maturation of the variety Kensington Pride in NT, Australia, counting from the date that the flowering bud is visible, or to 1300 units from the date when inflorescences have about two-thirds of open flowers (https://nt.gov.au/industry/agriculture/food-crops-plants-and-quarantine/fruit-crops/mango) (Fig. 7a).

Online heat sum calculators are available in some regions that utilise temperature data from a local government weather station, although better results will be obtained from use of an on-farm temperature record. Temperature is typically monitored using either a thermistor or a thermocouple, housed

© Burleigh Dodds Science Publishing Limited, 2018. All rights reserved.

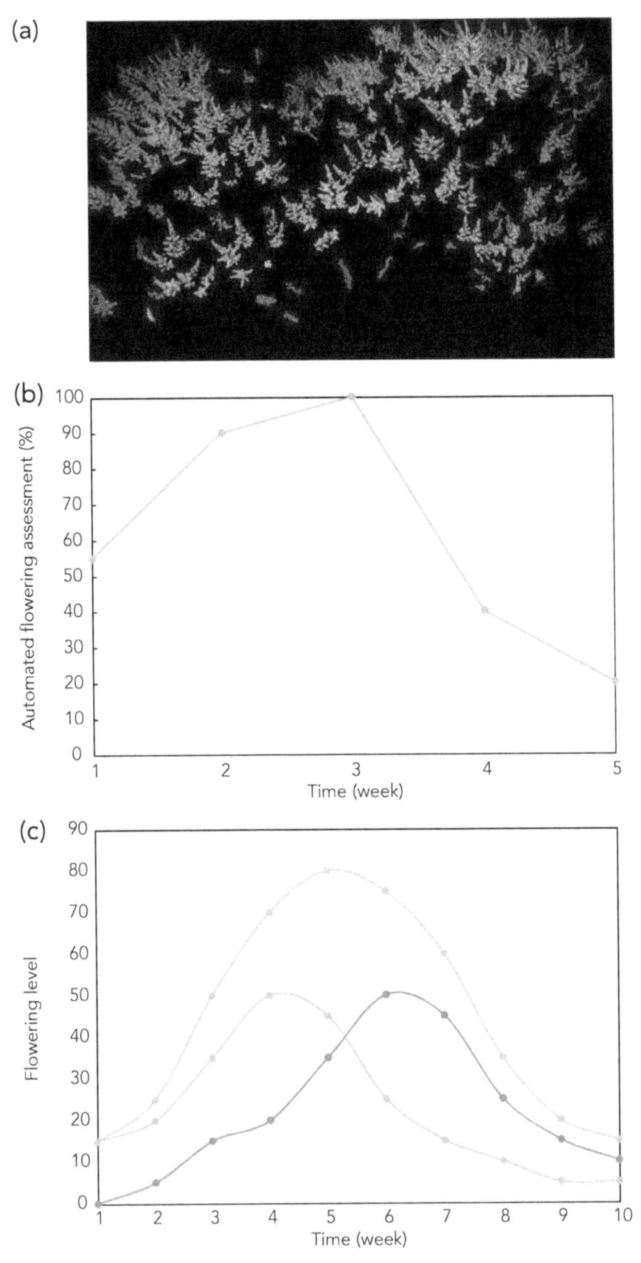

Figure 6 (a) Mango flowering at 'Christmas tree' stage - image segmented for inflorescences, (b) time course of inflorescence-associated pixels in canopy images, (c) artificial data: flower pixel count of two sequential flower events and associated total of inflorescence-associated pixels in canopy image. *Source* CQUniversity.

© Burleigh Dodds Science Publishing Limited, 2018. All rights reserved.

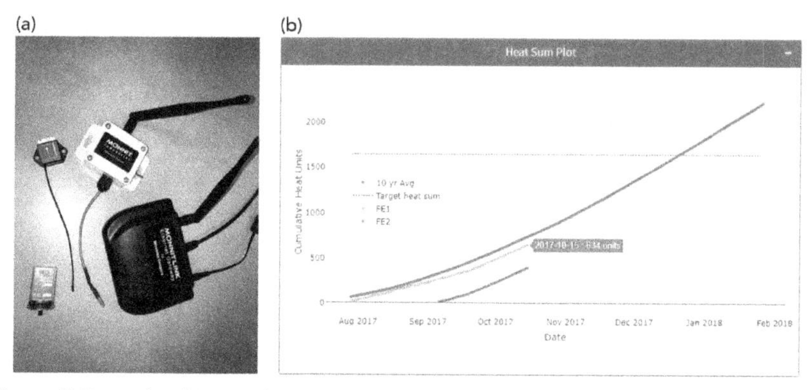

Figure 7 Example of (a) wireless temperature loggers and a receiving base station and (b) heat sums for a season (calculated for two flowering events, FE1 and FE2, for the current season temperature record, and for FE1 using the 10 year average temperature data). *Source* CQUniversity.

in a mini-Stevenson screen (i.e. a ventilated, shaded structure) (Fig. 7b,c). A thermistor is a temperature-sensitive resistor, while a thermocouple generates a voltage proportional to the temperature. Thermistors are typically used in horticultural applications, for which drift with temperature variation is greater than for thermocouples, but simpler electronic circuitry supports lower cost. A large number of commercially available on-farm weather stations or simple outdoor temperature loggers exist, including those able to broadcast wirelessly (radio, Bluetooth, Wi-Fi or phone networks) to a central station, linking via an Internet gateway to a cloud-based data centre for ease of data viewing (e.g. www.monnit.com). Utilising such data in decision support is considered in a later section.

2.3 Monitoring size and external appearance

A change in fruit shape occurs as fruit mature for many, but not all, varieties, typically seen as a 'filling' of the fruit shoulders (i.e. an increase in fruit thickness rather than width or length). This visual clue can be used in gauging crop harvest maturity and in selective harvest of more mature fruit.

The size distribution of fruit on tree, pre-harvest, can be useful to support tray insert size purchase and marketing decisions. For some fruits (e.g. cherries, citrus, apples, kiwifruit; Li et al., 2015; Green et al., 1990), the rate of change of fruit size on tree is used as an index of harvest maturity. Fruit dimensions (length, width and thickness) can be related to fruit weight (e.g. for 'Chok Anan' mango, R^2 = 0.96) (Spreer and Müller, 2011; Anderson et al., 2017; see also Fig. 9a). The rate of increase of mango fruit weight is decreasing but not plateaued by the time of harvest maturity (Fig. 8b).

© Burleigh Dodds Science Publishing Limited, 2018. All rights reserved.

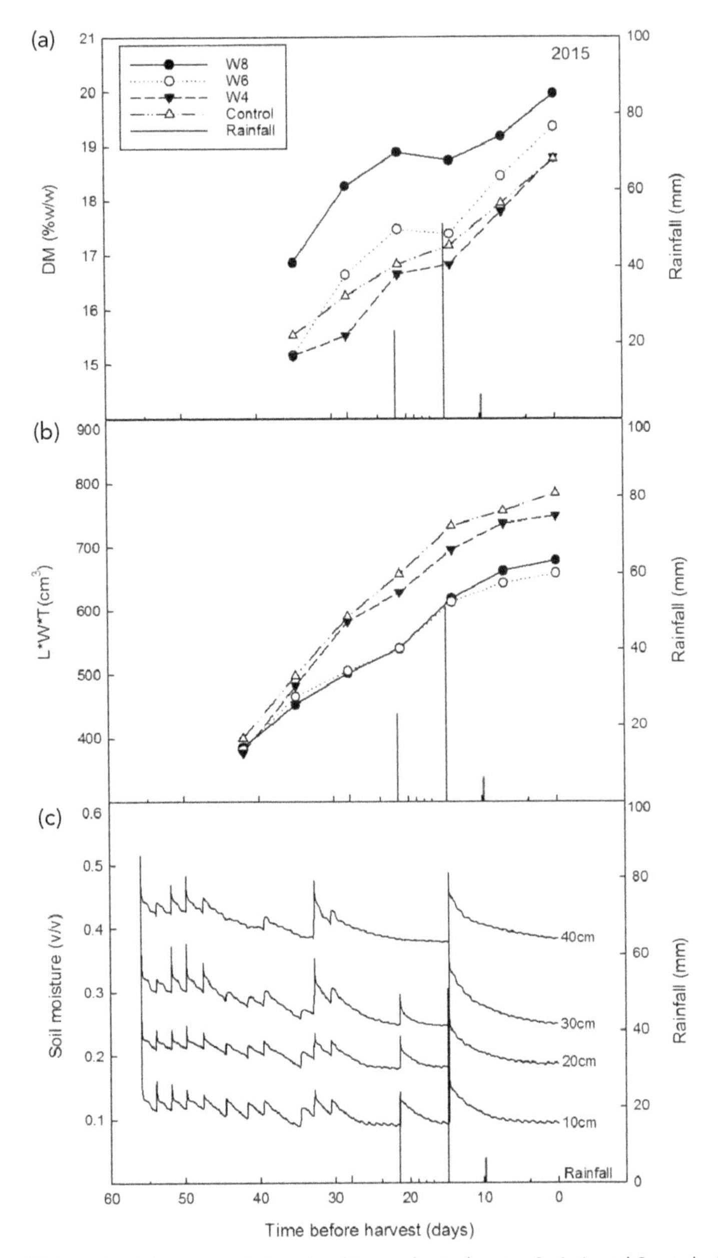

Figure 8 Water denial treatments involved trees denied water 0, 4, 6 and 8 weeks before harvest. (a) Weekly NIR DM of 30 fruits per treatment measured from five weeks before harvest, (b) L*W*T (cm^3) collected on the same fruit from six weeks before harvest, (c) soil moisture (v/v) at four soil depths from a probe in the 4-week denial treatment installed eight weeks before harvest and rainfall (mm) measured daily. Soil moisture values for different depths have been offset for ease of viewing. *Source* Anderson et al., 2017.

© Burleigh Dodds Science Publishing Limited, 2018. All rights reserved.

Mango fruit size can be estimated manually using callipers. This measurement requires a level of operator attention, particularly when manual transcription of results is required. Because of this requirement, infield fruit size measurements are rarely made in commercial mango production, and when undertaken, the labour requirement results in a small number of measurements, decreasing the statistical value of the information. Some digital versions of these measurement devices allow for transfer of data to a connected tablet, for example, Green et al. (1990); Gus P/L (http://www.gusstoday.com/datalogger.html). Other systems that allow continuous estimation of fruit size are available (e.g. Morandi et al., 2007; Phytek. http://www.phytech.com/); however, these systems require installation on a single piece of fruit, with equipment costs limiting multiple installations.

Vision-based systems are used for the estimation of fruit size in grading systems, under controlled lighting and background, typically estimating lineal dimensions to 1.0 mm, for example, Sadegaonkar and Wagh (2013) and Dhameliya et al. (2016); also see review by Moreda et al. (2009). A field machine vision-based fruit sizing (RMSE 3 mm) was described by Koirala et al. (2017), based on manual collection of images of fruit against an A4-sized blue background with scale. Images were collected using a mobile phone, with calculation of fruit weight based on the allometric relation with fruit minor and major axis dimensions, with information relayed with geolocation information to a cloud-based information system (Fig. 9).

Alternatively, infield fruit sizing information can be extracted from high-resolution images of whole canopies, from a tractor-mounted imaging system which might also do duty in flowering and fruit load assessment. In this approach only a representative number of fruit needs to be assessed, not all fruit, and thus analysis can be restricted to non-occluded and well-positioned fruit. Fruit size calculation requires depth information, from stereo vision or time-of-flight (TOF) laser imaging systems (e.g. Moreda et al., 2009; Wang et al., 2016). The TOF imaging systems typically use a wavelength of around 830 nm, so infield use is restricted to night time, with LED flood lighting.

2.4 Monitoring internal colour

Colour cards, for the estimation of flesh colour, are a useful aid to standardisation of human assessment - although these cards do fade and appear different with variation in lighting (Fig. 10a). Consistent measurement of colour can be achieved with commercial colorimeters (Fig. 10b); however, the cost of such units and the need for destructive sampling of fruit place a barrier to their use in farm production. There is a market need for a lower cost colorimeter which assesses a reasonable surface area (e.g. 20 cm^2) for the fruit industry.

Visible short-wave spectroscopy of intact fruit has also been used in the non-invasive estimation of flesh colour, using an optical geometry that ensures

© Burleigh Dodds Science Publishing Limited, 2018. All rights reserved.

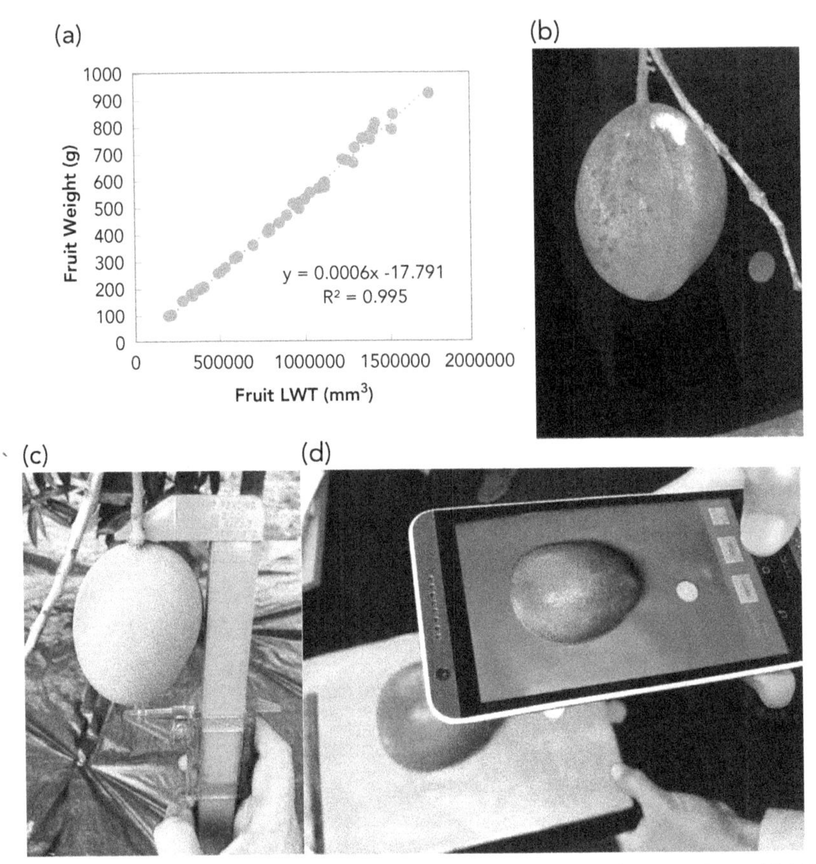

Figure 9 (a) Allometric relationship of fruit lineal dimensions with fruit weight and (b) CIE Lab b channel image of infield fruit against a blue background, (c) measurement of fruit dimensions infield using callipers, (d) measurement of fruit dimensions using phone app. *Source* CQUniversity.

passage of light through flesh (Subedi et al., 2007). Hand-held spectrometers allow measurement of fruit infield. However, the calibration between spectra and flesh colour requires adjustment between growing districts, likely due to change in skin properties with growing district (Subedi et al., 2007). The technique involves passage of light through skin and flesh, and changes in skin properties can impact the assessment of flesh colour.

2.5 Monitoring DM

Fruit DM is typically assessed by weight loss of a tissue sample following 24–48 h of oven drying, typically at 65°C. Domestic dehydrator units can be used, with use of a three decimal place (i.e. 1-mg resolution) balance recommended.

© Burleigh Dodds Science Publishing Limited, 2018. All rights reserved.

Microwave drying can be used, but care is required to achieve a consistent level of drying without loss of volatiles or charring.

A non-invasive technique allows repeated monitoring of individual fruit on tree, removing the sampling error associated with destructive sampling. The use of short-wave near-infrared spectroscopy for non-invasive assessment of DM in intact mango has been advocated for several decades. For example, Guthrie and Walsh (1997) reported a multiple linear regression model to

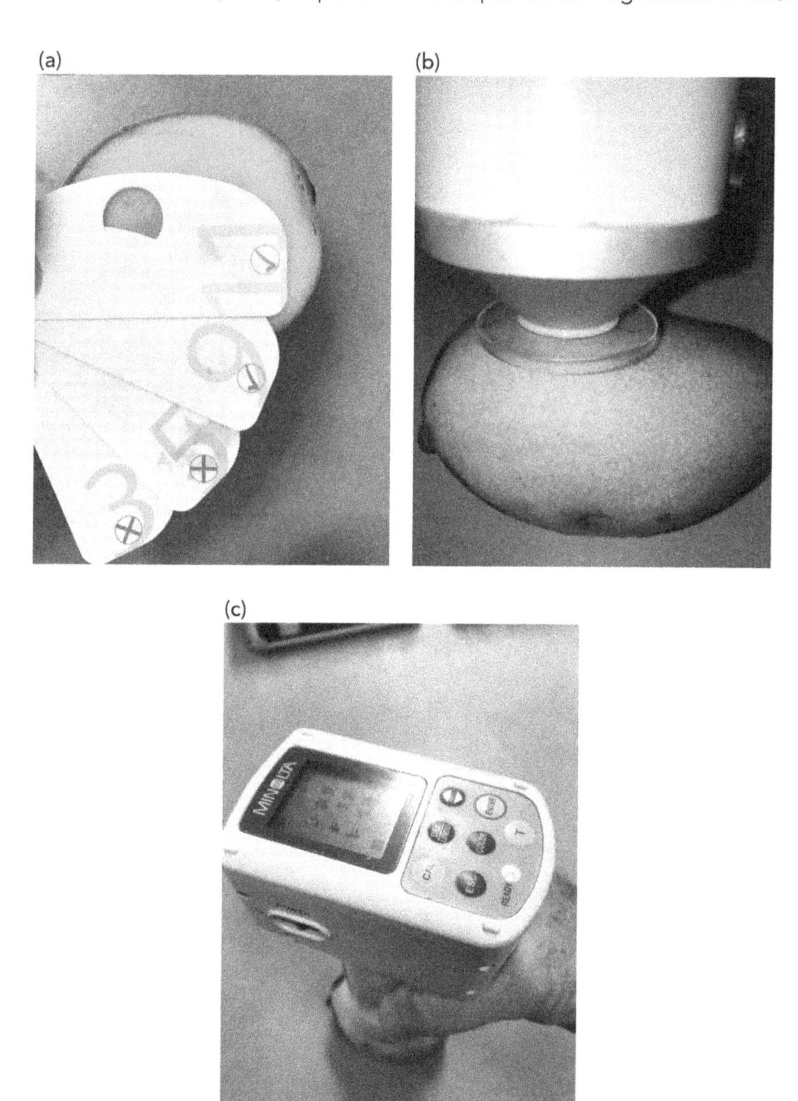

Figure 10 (a) Colour cards used in the assessment of colour of flesh of cut cheeks of mango fruit; (b) side view of colorimeter (Minolta CR400) with internal lighting for consistent estimation of colour; (c) top view of colorimeter. *Source* CQUniversity.

© Burleigh Dodds Science Publishing Limited, 2018. All rights reserved.

predict 'Kensington Pride' mango fruit DM with R^2 = 0.96, and RMSEP = 0.79. Saranwong et al. (2004) reported the use of a PLS model to predict DM in a validation set of hard green 'Mahajanka' mangoes, with R^2 = 0.92, SEP = 0.41, bias = 0.07, while Subedi et al. (2007) reported that a multi-cultivar mango model predicted independent populations of fruit with acceptable accuracy (R^2 = 0.79, RMSEP = 0.97).

This technique relies on the absorption of light, with absorbance bands around 740, 840 and 960 nm, associated with overtones of O-H bond stretching, and at 910 nm associated with an overtone of C-H bond stretching. In essence, the technique is assessing DM in terms of water, sugar and starch content. In practice, calibrations (partial least squares regression) between spectra of intact fruit (typically using the wavelength range 720-975 nm) and oven DM values routinely achieve a R^2 = 0.85 and a root mean square of error of prediction (RMSEP) = 0.7%DM (Fig. 11).

2.6 Manipulating DM

Given the ability to monitor fruit DM, the grower also requires tools to manipulate fruit to meet market DM specifications. Agronomic treatments that alter carbohydrate or water allocation to the fruit are candidate treatments. For example, Yeshitela et al. (2004) reported manual and chemical fruit thinning to result in increased SSC in ripened fruit (15.1-16.3° cf. control 13.7 % SSC). Simmons et al. (1998) reported that girdling of mango branches eight weeks following flowering to achieve a single fruit per girdled branch impacted both fruit weight (441, 363, 533 and 697 g/fruit for control, 30, 60 and 120 leaves/

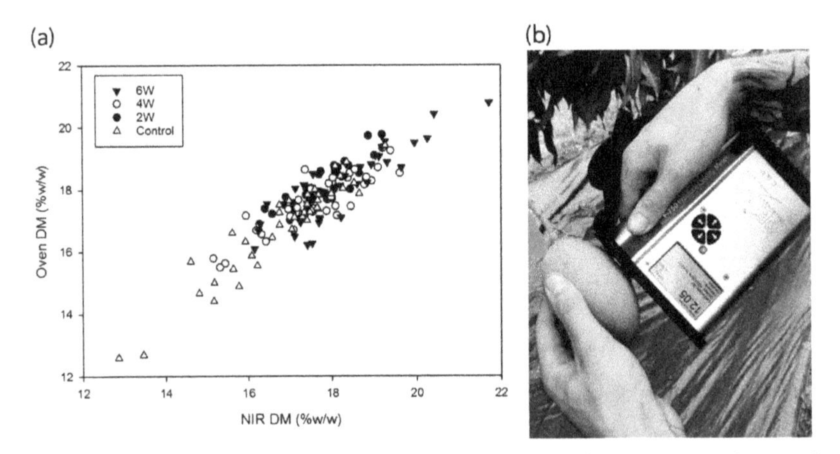

Figure 11 (a) Oven and NIR measurements of DM of fruit for trees denied water for zero, two, four and six weeks before harvest. (b) Handheld short-wave near-infrared spectrometer in use in a mango orchard. *Source* CQUniversity.

© Burleigh Dodds Science Publishing Limited, 2018. All rights reserved.

fruit treatments) and fruit DM (14.4, 16.4, 16.3, 14.6%, respectively). Zhao et al. (2013) reported fruit bagging to result in increased fruit carbohydrate content (e.g. 114.6 and 99.5 mg total sugars/g FW, in bagged and non-bagged fruit, respectively). Fruit carbohydrate concentration can be increased using deficit irrigation (DI), partial root-zone drying (PRD) and regulated deficit irrigation (RDI), for example, as reported for kiwi fruit (Miller et al., 1998), apple (Mpelasoka et al., 2001) and musk melon (Long et al., 2006). In general, fruit size is decreased if the water deficit is applied early in fruit development results, while storage reserve levels are increased if the deficit is imposed in late development. For mango fruit, Simmons et al. (1998) reported denial of irrigation water for 56 days following panicle emergence, 56 days before harvest and 14 days before harvest to result in harvest DM of 17.4, 14.4 and 12.7%, respectively. Nagle et al. (2010) also reported fruit DM to be increased when water was denied from flowering (24.6% compared to control at 21.4%), extending the result of Spreer et al. (2007). These treatments have also been explored by Anderson et al. (2017), who noted increased DM with thinning and water denial, but not trunk girdling (possibly too mild a girdling treatment) or fruit bagging (Table 1).

Of course, irrigation denial can only be effective in the absence of rain. In the example provided (from Anderson et al., 2017) in Fig. 12, rainfall events of approximately 20 and 50 mm at 23 and 16 days before harvest caused a temporary decrease in the rate of DM accumulation.

In summary, there is motive to increase DM to improve ripened soluble solids level and eating experience, requiring instrumentation to monitor fruit DM infield and agronomic practices to manipulate DM in the growing crop.

Table 1 DM of fruit at harvest for a number of treatments over several years and locations (Anderson et al., 2017). Water denial periods refer to the number of weeks before harvest that irrigation was discontinued

	Harvest DM%				
Treatment	Trial 1	Trial 2	Trial 3	Trial 4	Trial 5
Control	15.5	17.7	18.8	17.1	16.5
Thinning	17.0				
Trunk girdling					
Fruit bagging					
Water denial - 2 weeks					17.6
Water denial - 4 weeks		19.4	18.8		17.2
Water denial - 6 weeks			19.4		18.3
Water denial - 8 weeks		19.4	20.2		
Water denial - 10 weeks				18.5	

© Burleigh Dodds Science Publishing Limited, 2018. All rights reserved.

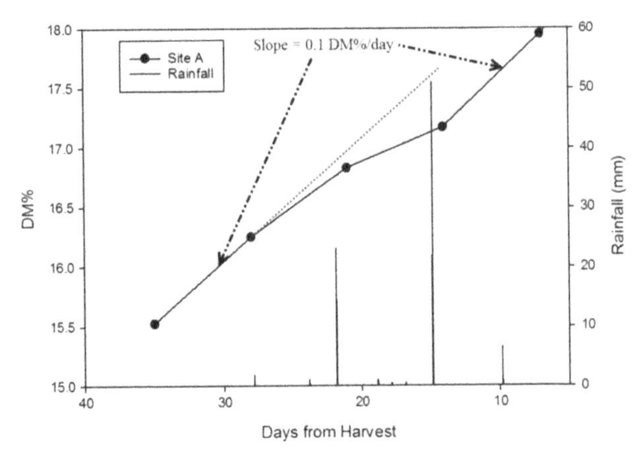

Figure 12 Time course of DM accumulation and rainfall. Modified from Anderson et al. (2017).

3 Monitoring quantity

3.1 On-tree monitoring

A rule of thumb best practice for estimation of orchard fruit load is to (manually) count all fruit on every 20th tree. Thus, in a block of 1000 trees, 50 trees would be assessed. Of course, the number of trees (n) required for a reliable estimate depends on the variation [standard deviation (SD)] on fruit load per tree - if there was no variability, a single tree could be assessed (Eqn. 1).

$$n = \left(t.\frac{SD}{e}\right)^2 \quad (1)$$

where t is the t statistic for a given probability (e.g. 1.96 for a 95% probability, $n > 30$) and e is the desired level of accuracy.

For example, the orchard described by Stein et al. (2016) had tree fruit count per tree ranging from 1 to 442, with an average of 102 and SD of 81 fruit/tree. Thus, if an error on the estimate of mean fruit count per tree of 10 fruit per tree is accepted, the number of trees to be assessed is $(1.96 \times 81/10)^2 = 252$. Note that this number is independent of population size, but is dependent on a reliable estimate of SD. Thus categorising trees in classes based on fruit load can reduce sampling effort. For example, if in the example, trees could be categorised to three types, each with SD on fruit load of 20, then n = (1.96 × 20/10)2 = 16 trees should be sampled in each category or 48 trees in total. Measures of trunk diameter, canopy volume (by LiDAR, light detection and ranging) and canopy health (by spectral indices such as NDVI, as suggested by Robson et al., 2016) are attributes of potential value for stratification of

© Burleigh Dodds Science Publishing Limited, 2018. All rights reserved.

Figure 13 Machine vision estimation of fruit load infield is a developing technology. (a) Farm utility mounted machine vision rig used in flowering and fruit and (b) night imaged mango canopy. *Source* CQUniversity.

fields to regions that are more homogenous in maturity/DM, thus requiring less sampling.

Manual counting is tedious and requires continuous concentration. The counter must adopt a pattern - for example, counting from a distinguishable branch and moving in one direction around the tree. Counting fruit load on more than every 20th tree becomes an unrealistic task, especially for larger trees. Fruit load estimation using machine vision is yet to see commercial adoption, but capacity is rapidly advancing (Fig. 13). For example, Payne and Walsh (2014) reviewed the use of various technologies (colour, thermal and colour imaging, LiDAR) and Payne et al. (2014) reported on the use of night colour imagery to improve contrast between fruit and background. Quereshi et al. (2016) employed this imagery with machine vision features of K-nearest neighbour pixel classification and contour segmentation and a method based on super-pixel over-segmentation and classification using support vector machines. More recently, Stein et al. (2016) presented a method to localise every piece of fruit in a mango orchard using tracking of fruit between multiple images ('multi-view') of the canopy to locate fruit occluded by foliage or other fruit in a single frame image of the tree, coupled with use of a faster regional convolutional neural network (R-CNN) detector for fruit detection and a

© Burleigh Dodds Science Publishing Limited, 2018. All rights reserved.

LiDAR-generated mask for each canopy, with all fruit assigned to a tree. The 'multi-view' is a form of stereovision, allowing each fruit to be assigned a 3D location.

3.2 In-packhouse monitoring

In almost all supply chains, mango fruit will pass through a packhouse after harvest. The packhouse allows for functions such as cleaning and application of protective treatments, as well as for categorisation and quantification of fruit (Fig. 14). Fruit are typically singulated onto a cup conveyor and are sorted on

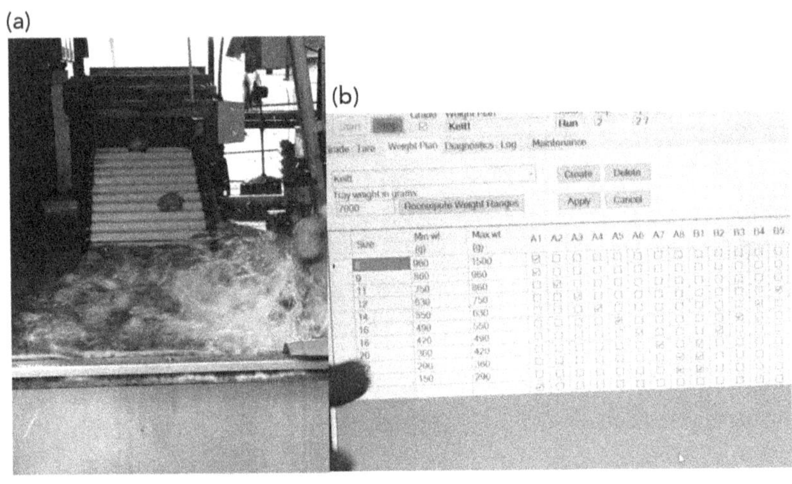

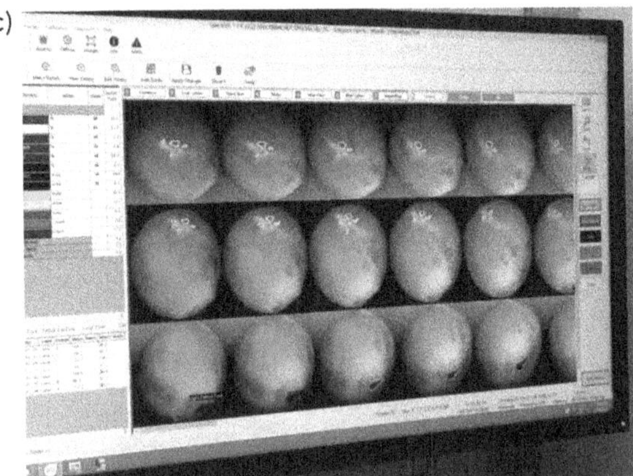

Figure 14 (a) Water dump of fruit from field bins, with fruit moved up a roller bed to a singulator that places fruit into cups that are weighed before tipping at designated packing stations, (b) operator input to a weight only electronic grader, assigning fruit-to-weight grade ranges and to pack out points on the grader, with weight ranges matching to sizes (8–22 pieces of fruit packed to each 7.5 kg tray), (c) defect sorting of mango fruit on a packline using machine vision.

© Burleigh Dodds Science Publishing Limited, 2018. All rights reserved.

the basis of weight using a load cell (situated under the conveyor, weighing every cup and fruit), with human sorting of visual defects (e.g. skin rub, sun bleach, lenticel damage). Equipment is also available for machine vision-based sorting of fruit on colour and external defects.

4 Monitoring ripeness

Following grading, mango fruit consignments are typically ripened to a determined level before distribution. Measurements useful to monitoring and control of the ripening process include (i) temperature, (ii) skin colour, (iii) firmness and (iv) ethylene level. For apple in storage, commercially available systems such as Harvest Watch (www.harvestwatch.net/) allow detection of the onset of storage disorders through measurement of chlorophyll fluorescence; however, this technology has not been applied to mango given that the fruit is not stored for long periods.

4.1 Monitoring temperature

Fruit and store temperatures are typically logged using a thermistor, but infrared thermometers (which measure infrared wavelengths around 5 μm emitted from an object) have found favour, being non-contact and available in imaging as well as point modes. However, the technique is a surface measurement. If fruit core temperature is required, a thermocouple must be inserted into the fruit. An interesting development is that of logging and cloud-based cold chain decision support systems (e.g. XSense, www.bt9-tech.com).

4.2 Monitoring colour

Skin pigmentation can be visually assessed, with the use of colour charts to help maintain consistency of assessment (Fig. 15a). An instrumentation-based assessment of skin chlorophyll content is possible, for example, the handheld DA meter (Fig. 15b) uses alternating LED illumination at 680 and 720 nm to calculate an index related to fruit chlorophyll content. This index decreases during fruit ripening and could be used as a ripening stage index (Fig. 15c), although use of the instrument at a constant temperature is recommended, given shift in LED peak output wavelength with temperature (Hayes et al., 2017).

4.3 Monitoring firmness

Flesh firmness is currently routinely assessed using hand feel, with some attempt to standardise assessment by use of reference latex balls of different firmness (Fig. 16a). A penetrometer is recommended for more accurate, although destructive, measurement. However, handheld penetrometers are subject to operator variation, for example, in speed of insertion. To reduce this variation, use of a motorised stand is recommended (Fig. 16b).

© Burleigh Dodds Science Publishing Limited, 2018. All rights reserved.

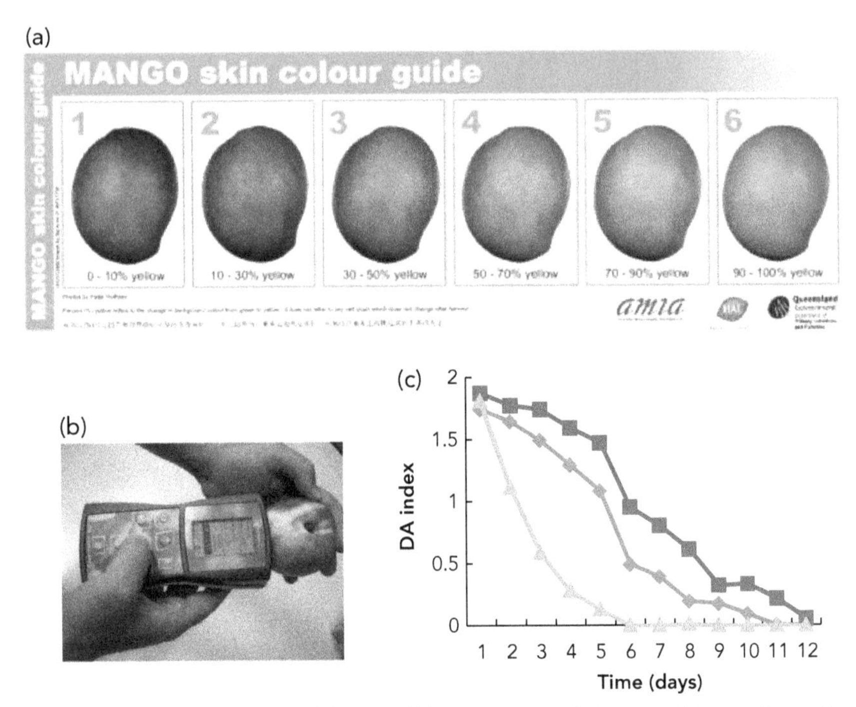

Figure 15 (a) Colour chart of fruit at different stages of ripeness (*Source* Australian Mango Industry Association), (b) DA meter and (c) change in DA values of three ripening fruit from a common lot (*Source* CQUniversity).

In commercial practice, the destructive nature of a penetrometer limits sampling effort. Non-destructive alternatives exist, although not adopted by industry to date. The extent of deformation under a set load can be assessed, mimicking the mechanism of hand feel (e.g. the Agrosta14 (http://www.agro-technologies.com/ang/produits/crisp_fruits.htm). Fruit acoustic properties also change as flesh firmness (and Young's modulus of the material) changes. This can be assessed by either a change in the velocity of a pressure (acoustic) wave travelling through the fruit or by the resonant frequency of the fruit (Fig. 16c) (e.g. Subedi and Walsh, 2008). Both systems rely on a sharp but gentle tap of the fruit, producing a pressure wave within the fruit. As this pressure wave reaches the fruit surface, it causes vibration in the surrounding air, detectable by an audio microphone. There will be a dominant (resonant) frequency. To estimate velocity of the pressure wave, microphones are placed at two distances from the impact point.

4.4 Monitoring ethylene and CO_2

Mango fruit benefit from a controlled ripening process, with the estimation of ethylene and CO_2 desirable. The concentration of these gases can be

© Burleigh Dodds Science Publishing Limited, 2018. All rights reserved.

(a)

Stage	Description
Hard	No give under strong thumb pressure
Rubbery	Slight give under strong thumb pressure
Sprung	fruit deforms by 2-3 mm under moderate thumb pressure
Firm soft	fruit deforms by 2-3 mm under slight thumb pressure
Soft	whole fruit deforms with slight hand pressure

(b) (c)

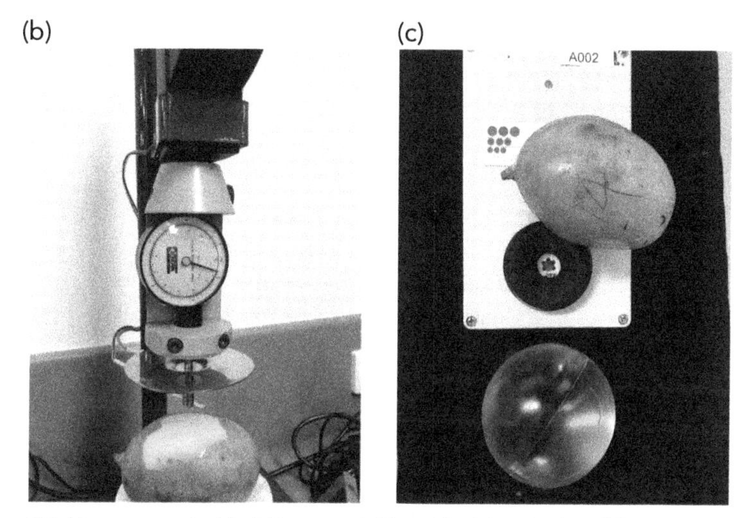

Figure 16 Measurement of fruit firmness: (a) hand feel description, (b) penetrometer on motorised stand and (c) acoustic frequency device.

monitored using a colorimetric gas detection tube (Diels-Alder reaction, Dewar et al., 1986; e.g. the Kitagawa ethylene detector tube, Fig. 17a). Typically, a gas aspirating pump is used to draw air through the tube. Similar devices exist for CO_2. Alternatively, an electrochemical cell can be used in the estimation of ethylene, and infrared absorption for the estimation of CO_2, for example, the Felix Instruments F960, Fig. 17b.

5 Decision support systems

A family farm might have fewer than ten blocks, with each block of fewer than 1000 trees, involving several varieties to give a spread of harvest windows and marketing opportunities. This operation is small enough for the farm manager to keep a mental track of individual block maturity and other issues. In comparison, a corporate farming group may be dealing with a single cultivar under Plant

© Burleigh Dodds Science Publishing Limited, 2018. All rights reserved.

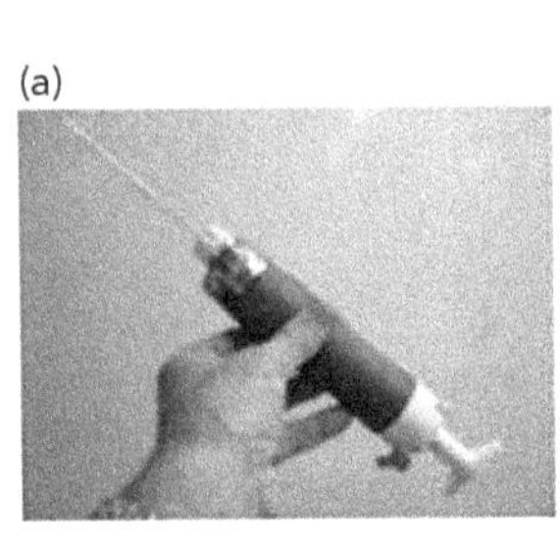

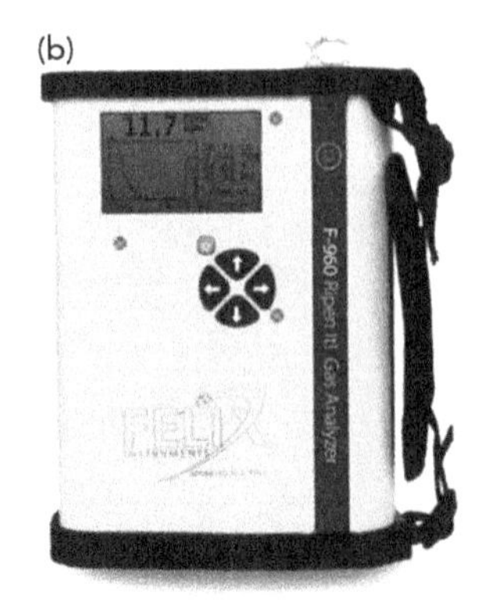

Figure 17 Ethylene and CO_2 gas concentration measurement: (a) colorimetric detection (Kitsgawa detection tube) and (b) electrochemical – IR (Felix F-960). (*Source* http://www.kitagawa-america.com/admin/efiles/PR11346.pdf; https://felixinstruments.com/).

Variety Rights, with a number of large operations dispersed across different geographic locations to provide an extended harvest window. Each operation can have >70 blocks of 1000 trees each, with harvest at a given location occurring over a tight window. In this circumstance, there is a clear benefit for the use of decision support aids.

As described earlier, tools to measure a range of parameters relevant to mango crop quality and quantity exist. However, for a farm manager to utilise this range of information in a timely fashion, the information must be 'digested' and presented in a visual format that enables quick comprehension. A good DSS will present information in a simplified and easily interpretable way, assisting management decisions. An example DSS for mango production is the 'Mango maturity' Web app (www.fruitmaps.info; for a demonstration version, enter 'demo' in all security checks) (Walsh, 2017), described in the following text.

The first requirement of a farm management app is a tool for the user to enter management boundaries, that is, tree blocks, with display to a satellite or aerial map. More sophisticated management may require several layers of boundaries, for example, blocks defined by harvest units, planting age, irrigation zones or soil type.

Ideally all data inputs should be automated. For example, to simplify the estimation of heat units, a temperature sensor installed on farm can log data to the Web app. Alternatively, temperature data could be sourced from the Web, either from a nearby publically available weather station or by a payment service offering interpolated data for the farm site. Given input of time of flowering events, a graphic that conveys maturation information easily to the grower is a graph of accumulated heat units to date for the

© Burleigh Dodds Science Publishing Limited, 2018. All rights reserved.

current year together with a prediction of the date of fruit maturation based on a projection of the current year's temperature, relative to the historical temperature average and the mango cultivar (Fig. 7b).

Fruit count data (estimated after stone hardening stage, when fruit drop decreases) associated with each flowering event and geolocation records can be uploaded to the Web app for calculation of fruit load per block given information on number of trees per block. Fruit number associated with a given flowering event can be summed across orchard blocks to provide expected farm fruit numbers by harvest date. This system relies on manual estimates at present, but there is potential to have information input from machine vision estimates in the future.

The Web app utilises fruit DM readings from the F-750 Produce Quality Meter, with its geolocation records. A minimum input of a survey of fruit across a block, followed by one follow-up measurement to allow an estimate of a rate of DM increase, is suggested. The 'Farm DM' tab displays a map showing location of individual records, with colour change if the measurement exceeds the user-selected target value (Fig. 18). The block colour changes when the sampled fruit values fit the '% of fruit above target' criterion.

The 'Block DM detail' tab displays data of a single block, with current block mean DM, rate of increase of DM across time and the estimated date to achieve the user-defined criteria (for example 90% of fruit above 15% DM) given that rate of increase (Fig. 19). Another tab displays a list of all blocks of the farm, ordered by DM level and recommended date of harvest (i.e. date at which user set criterion on DM level is met), from the user set rate of increase. Thus this example of a DSS offers support around the decision to harvest (i.e. estimation of harvest maturity) based on heat sums, DM and crop load information.

6 Future trends and conclusion

Bill Gates of Microsoft is attributed with saying 'We always overestimate the change that will occur in the next two years and underestimate the change that will occur in the next ten'. Over the next decade, technology and social trends from the broader society will sweep into the mango industry, in the form of satellite or drone-based inventories of the crop, autonomous vehicles and autonomous harvesting, variable rate spraying and fertiliser driven by better understanding of spatial variability or yet unforeseen applications.

Likely nascent applications include:

- Improvement of heat sum models from input of a daily average of maximum and minimum temperature to more complete temperature profile data and sunshine hours, as for apple (Li et al., 2015).

© Burleigh Dodds Science Publishing Limited, 2018. All rights reserved.

Figure 18 DSS Web app for mango maturity, with display of fruit DM by location, colour coded for individual fruit above (blue) and below (red) a specification (15% DM), and block performance relative to a specification (90% of fruit above 15% DM) (fail in red, just passed, yellow).

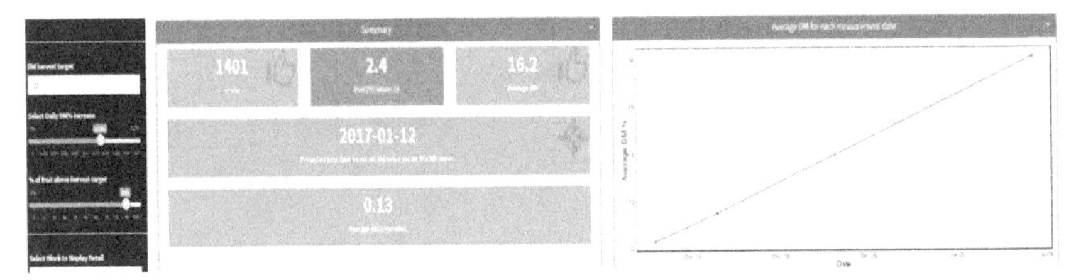

Figure 19 DSS Web app for mango maturity, with display of individual block data, including current mean DM, rate of weekly increase in DM and estimated date to achieve the criterion set in the left panel. User feedback on number of fruit sampled is provided, based on SD of data (thumbs up/down symbol).

- Machine vision record of early flowering trees to enable variable rate spraying, with sprays turned off for non-flowering trees.
- Machine vision record of early flowering trees to inform selective harvest, with earlier flowering trees having fruit that will mature earlier.
- Machine vision guided selective fruit harvest.
- Matching of predicted harvest maturities to markets.

7 Where to look for further information

The International Society for Horticultural Science's 'Mango Symposium' is held every second year, with presentations published in the *Acta Horticulturae* series

© Burleigh Dodds Science Publishing Limited, 2018. All rights reserved.

(see http://www.ishs.org/mango). This Symposium and the associated working group is an excellent source of expertise. Scientific publications relevant to the material of the current chapter is found in a wide range of journals, but key publications are *Computers and Electronics in Agriculture*, *Precision Agriculture*, *Postharvest Biology* and Technology and *Scienia Horticulturae*.

8 Acknowledgements

We acknowledge input of team members A. Koirala, P. Subedi and N. Anderson and of M. Matzner of Acacia Hills Farms and support of Horticulture Industry Australia project ST15005. We thank D. Swain for assistance in initiating the RStudio DSS software.

9 References

Anderson, A., Subedi, P. P. and Walsh, K. B. (2017). Manipulation of mango fruit dry matter content to improve eating quality. *Scientia Horticulturae* 226: 316–21.

Campbell, T. (2015). *Mango Quality Standards*. MG14504 final report. Horticulture Innovation, Australia.

Dewar, M. J., Olivella, S. and Stewart, J. J. (1986). Mechanism of the Diels-Alder reaction: Reactions of butadiene with ethylene and cyanoethylenes. *Journal of the American Chemical Society* 108: 5771–9.

Dhameliya, S., Kakadiya, J. and Savant R. (2016). Volume estimation of mango. *International Journal of Computer Applications* 143(12): 11–16. ISSN: 0975-8887. http://www.sciencedirect.com/science/article/pii/S0021863498902852.

Diehl, D. C., Nicole, L., Sloan, N. L., Bruhn, C. M, Simonne, A. H., Brecht, J. K. and Mitcham, E. J. (2013). Exploring produce industry attitudes: Relationships between postharvest handling, fruit flavor, and consumer purchasing. *HortTechnology* 23: 642–50.

Green, A. E., McAneney, K. J. and Astill, M. S. (1990) An instrument for measuring kiwifruit size. *New Zealand Journal of Crop and Horticultural Science* 18: 115–20, doi:10.10 80/01140671.1990.10428080.

Guthrie, J. and Walsh, K. B. (1997) Non-invasive assessment of pineapple and mango fruit quality using near infra-red spectroscopy. *Australian Journal of Experimental Agriculture* 37: 253–63.

Hayes, C. J., Walsh, K. B. and Greensill, C. V. (2017) Light-emitting diodes as light sources for spectroscopy: Senistivity to temperature. *Journal of Near Infrared Spectroscopy*. http://journals.sagepub.com/doi/pdf/10.1177/0967033517736164#articleCitation DownloadContainer.

Henriod R., Sole D., Wright C. and Campbell T. (2015) Mango Quality Standards by Terry Campbell. *Horticulture Industry Australia Project MG14504 Final Report*. HIA, Sydney, Australia.

Jordan, R. B., Seelye, R. J. and McGlone V. A. (2001) A sensory based alternative to Brix/ acid ratio. *Food Technology* 55(6): 36–44.

Koirala, A., Wang, Z., McCarthy, C. and Walsh, K. B. (2017). *An In-field Sizing Tool for Mango Fruit*. International mango Symposium, Nanning, China.

© Burleigh Dodds Science Publishing Limited, 2018. All rights reserved.

Ledger, S. Barker, L., Hofman, P., Campbell, J., Jones, V., Holmes, R. and Campbell, T. (2012). *Mango Ripening Manual*. Queensland Dept. of Agriculture, Fisheries and Forestry, Brisbane, Australia. https://static1.squarespace.com/static/53b0ef57e4b04ed3debabc4f/t/582a62939de4bb863459c7ef/1479172774823/Mango+Ripening+Manual.pdf. Accessed date: 13 January 2017.

Li, M., Chen, M., Zhang, Y., Fu, C., Xing, B., Li, W. and Yang, X. (2015) Apple fruit diameter and length estimation by using the thermal and sunshine hours approach and its application to the digital orchard management information system. *PLoS ONE* 10(4): e0120124. http://doi.org/10.1371/journal.pone.0120124

Long, R. L., Walsh, K. B., Midmore, D. J. and Rogers, G. (2006). Irrigation scheduling to increase muskmelon fruit biomass and soluble solids concentration. *HortScience* 41 (2): 367-9.

Miller, S., Smith, G., Boldingh, H. and Johansson, A. (1998) Effects of water stress on fruit quality attributes of kiwifruit, *Annals of Botany* 81(1): 73–81.

Mitcham, E. J. and McDonald, R. E. (1993) Respiration rate, internal atmosphere, and ethanol and acetaldehyde accumulation in heat-treated mango fruit. *Postharvest Biology and Technology* 3: 77–86.

Moreda, G. P., Ortiz-Cañavate, J., García-Ramos, F. J. and Ruiz-Altisent, M. (2009). Non-destructive technologies for fruit and vegetable size determination – a review. *Journal of Food Engineering* 92(2): 119–36. ISSN: 0260-8774.

Morandi, B., Manfrini, L., Zibordi, M., Massimo Noferini, M., Fiori, G. and Grappadelli, L. C. (2007). A Low-cost device for accurate and continuous measurements of fruit diameter. *HortScience* 42: 1380–2. http://hortsci.ashspublications.org/content/42/6/1380.full

Mpelasoka, B., Behboudian, M. and Green, S. (2001). Water use, yield and fruit quality of lysimeter-grown apple trees: Responses to deficit irrigation and to crop load. *Irrigation Science* 20(3): 107–13.

Nagle, M., Mahayothee, B., Rungpichayapichet, P., Janjai, S. and Müller, J. (2010) Effect of irrigation on near-infrared (NIR) based prediction of mango maturity, *Scientia Horticulturae* 125(4): 771–4.

Obenland, D., Collin, S., Mackey, B., Sievert, J., Fjeld, K. and Arpaia, M. L. (2009) Determinants of flavor acceptability during the maturation of navel oranges *Postharvest Biology and Technology* 52: 156–63 DOI:10.1016/j.postharvbio.2009.01.005

Payne, A. and Walsh, K. (2014) Chapter 16: Machine vision in estimation of crop yield. In: *Plant Image Analysis: Fundamentals and Applications* (S. Dutta-Gupta and Y. Ibaraki (Eds)). CRC Press, Taylor & Francis Group, pp. 330–72. http://www.crcpress.com/product/isbn/9781466583016

Payne, A., Walsh, K. B., Subedi, P. and Jarvis, D. (2014) Estimating mango crop yield using image analysis using fruit at 'stone hardening' stage and night time imaging. *Computers and Electronics in Agriculture* 100: 160–7.

Quereshi, W. S., Payne, A., Linker, R., Cohen, O and Dailey M. N. (2016) Machine vision for counting fruit on mango tree canopies. *Precision Agriculture* 18(2): 224–44. doi: 10.1007/s11119-016-9458-5

Robson, A. J., Petty, J., Joyce, D. C., Marques, J. R. and Hofman, P. J. (2016). High resolution remote sensing, GIS and Google Earth for avocado fruit quality mapping and tree number auditing. *Acta Horticulturae* 1130: 589–95.

© Burleigh Dodds Science Publishing Limited, 2018. All rights reserved.

Sadegaonkar, V. D. and Wagh, K. H. (2013). Quality inspection and grading of mangoes by computer vision and image analysis. *International Journal of Engineering Research and Applications* 3(5): 1208–12. ISSN: 2248-9622. www.ijera.com.

Saranwong, S, Sornsrivichai, J and Kawano, S 2004, Prediction of ripe-stage eating quality of mango fruit from its harvest quality measured nondestructively by near infrared spectroscopy. *Postharvest Biology and Technology* 31 (2): 137–45.

Simmons, S., Hofman, P., Whiley, A. and Hetherington, S. (1998). Effects of preharvest calcium sprays and fertilizers, leaf: Fruit ratios, and water stress on mango fruit quality. *ACIAR Proceedings* 81: 19–2

Spreer, W., Nagle, M., Neidhart, S., Carle, R., Ongprasert, S. and Müller, J. (2007). Effect of regulated deficit irrigation and partial rootzone drying on the quality of mango fruits (*Mangifera indica* L., cv.'Chok Anan'). *Agricultural Water Management* 88(1): 173–80.

Spreer, W. and Müller, J. (2011). Estimating the mass of mango fruit (Mangifera indica, cv. Chok Anan) from its geometric dimensions by optical measurement. *Computers and Electronics in Agriculture* 75(1): 125–31.

Stein, M., Bargoti, S. and Underwood, J. (2016). Image based mango fruit detection, localisation and yield estimation using multiple view geometry. *Sensors* 16: 1915.

Subedi, P., Walsh, K. B. and Owens, G. (2007). Prediction of mango eating quality at harvest. *Postharvest Biology and Technology* 43(3): 326–34.

Subedi, P. P and Walsh, K. B. (2008). Non - invasive measurement of fresh fruit firmness. *Journal of Postharvest Biology and Technology* 51: 297–304. doi:10.1016/j.postharvbio.2008.03.004

Walsh, K. B. (2015). Nondestructive assessment of fruit quality. In: *Advances in Postharvest Fruit and Vegetable Technology. CRC Press Series on Contemporary Food Engineering* (R. B.H. Wills and J. B. Golding (Eds)). Elsevier, CRC Press, Taylor & Francis Group, pp. 40–61.

Walsh, K. B. (2017) Fruit maturity: Developing an application to assist with the decision to pick. *Mango Matters* http://www.industry.mangoes.net.au/resource-collection/?tag=Dry+Matter Accessed date: 12 April 2017.

Wang, Z., Verma, B., Walsh, K. B. Subedi, P. and Koirala, A. 2016. Automated mango flowering assessment via refinement segmentation. In: *2016 International Conference on Image and Vision Computing New Zealand (IVCNZ).* IEEE Palmerston North, pp. 1-6. IEEE Catalog Number: CFP1667E-ART ISBN: 978-1-5090-2748-4 Online ISSN: 2151-2205, https://opus.lib.uts.edu.au/bitstream/10453/80091/1/proceedings%20IVCNZ%202016%20-%20front%20matter.pdf, doi:10.1109/IVCNZ.2016.7804426. http://ieeexplore.ieee.org/stamp/stamp.jsp?tp=&arnumber=7804426&isnumber=7804412.

Whiley, A. W., Hofman, P. J., Christiansen, H., Marques, R., Stubbings, B. and Whiley, D. G. 2006, Development of Pre and Postharvest Protocols for Production of Calypso Mango/Anthony Whiley. Final report/Horticulture Australia; HAL FR02049. Horticultural Australia, Sydney.

Yeshitela, T., Robbertse, P. and Fivas, J. (2004). Effects of fruit thinning on 'sensation'mango (Mangifera indica) trees with respect to fruit quantity, quality and tree phenology. *Experimental Agriculture* 40(4): 433–44.

Zhao, J. J., Wang, J. B., Zhang, X. C., Li, H. L. and Gao, Z. Y. (2013). Effect of bagging on the composition of carbohydrate, organic acid and carotenoid contents in mango fruit. *Acta Horticulturae* 992: 537–54.

© Burleigh Dodds Science Publishing Limited, 2018. All rights reserved.

Chapter 4

Advances in understanding and improving the nutraceutical properties of cranberries

Oliver Chen, Biofortis Research, Merieux NutriSciences and Tufts University, USA; and Eunice Mah, Biofortis Research, Merieux NutriSciences, USA

1 Introduction

American cranberry (*Vaccinium macrocarpon*) is commonly consumed as juice, dried fruit, or dietary supplement. It is mainly cultivated in the United States and Canada (Neto and Vinson, 2011). Native Americans used them as food, dye for rugs and blankets, and healing agent for wounds. On average, Americans eat 2.8 cranberries per day, equivalent to 0.25 mL of juice, making them an under-consumed fruit in the United States (Neto and Vinson, 2011). Cranberries are known for their tart taste and are generally consumed after being sweetened, for example, in the form of cranberry juice cocktails containing a variety of sweet juices, such as apples and grapes. Cranberry juice is perhaps best known for the prevention and treatment of urinary tract infections (UTIs). This health benefit has received a US FDA qualified health claim in 2020 (FDA, 2020). Nevertheless, the health benefits of cranberry products extend beyond the protection from infection in the urogenital tract to bacteria-related health conditions, such as dental and gastrointestinal health, and others, such as cardiometabolic and skin health, cognition, and cancer.

In this chapter, we first review clinical evidence supporting the health benefits of consumption of cranberry products, followed by a discussion on future directions in the elucidation of the mechanism of actions for health

http://dx.doi.org/10.19103/AS.2022.0101.16
© Burleigh Dodds Science Publishing Limited, 2022. All rights reserved.

benefits, as well as approaches to maximizing bioefficacy of cranberry-related foods/products. In this chapter, all discussed information is pertinent only to the American cranberry and related products but not European cranberry (*V. oxycoccos*).

2 Nutrient composition

Nutrition information on raw cranberries is presented in Table 1. Raw cranberries (1 cup, 100 g) have 46 kcal (mainly from carbohydrates), 87.3 g of water, and 3.6 g of fiber (Central, 2020). Cranberries are also a rich source of phytochemicals, including flavonoids (particularly flavan-3-ols), A-type proanthocyanidins (PACs), anthocyanins (cyanidin-3-arabinoside, cyanidin-3-galactoside, and cyanidin-3-glucoside), benzoic acid (quinic, citric, ellagic, and malic acids), ursolic acid, and triterpenoids. Due to the distinctive tart taste and astringency of phenolic compounds ascribed to their capability of binding saliva proteins, sweeteners are generally added to cranberry products, such as juices and dry fruit, to enhance palatability. The contents of anthocyanins, flavonols, and phenolic acids in raw cranberries can be found in the Phenol-Explorer (http://phenol-explorer.eu/) (Phenol-Explorer) and the USDA Database for the Flavonoid Content of Selected Foods (https://data.nal.usda.gov/dataset/usda-database-flavonoid-content-selected-foods-release-32-november-2015) (Bhagwat and Haytowitz, 2016). As reported in the Phenol-Explorer, in each subfamily, peonidin 3-O-galactoside, quercetin 3-O-galactoside, and benzoic acid are the most abundant with a mean content of 22.02, 10.81, and 48.10 mg/100 fresh weight, respectively. Total catechins in raw cranberry average 17 mg/100 g, with epicatechin being the most abundant (Harnly et al., 2006). PACs are the products of the polymerization of flavan-3-ols. A-type PACs (C2→O→C7 linkage between epicatechin units) are present in cranberries instead of the B-type PACs commonly found in most plant foods (Gu et al., 2003). As reported in the USDA Database for the Proanthocyanidin Content of Selected Foods (https://data.nal.usda.gov/dataset/usda-database-proanthocyanidin-content-selected-foods-release-2-2015) (Bhagwat and Haytowitz, 2015), the most abundant PACs in raw cranberries are polymers (>10 flavan-3-ols unit) at 217.6 mg per 100 g, followed by 4–6 mers at 56.84 mg/100 g. The average concentration of PACs is 231 mg/L in cranberry juice (Gu et al., 2004). PACs with a degree of polymerization larger than four are not absorbable in the small intestine because their large molecular size prohibits them from passing the gut barrier (Ou et al., 2012).

The variability in bioactive compounds among cranberry products partially explains the non-reproducibility of clinical evidence between studies on cranberries and UTI and other health indications. Among constituents in cranberries, A-type PACs have been recognized for UTI benefits and thus

© Burleigh Dodds Science Publishing Limited, 2022. All rights reserved.

Table 1 Nutrient composition in raw cranberries (100 g)[a]

Kcal	46
Water (g)	87.3
Carbohydrate (g)	12
Fat (g)	0.13
Protein (g)	0.46
Sugar (g)	4.27
Fiber (g)	3.6
Vitamin C (mg)	14
alpha-Tocopherol (mg)	1.32
beta-Carotene (μg)	38
Lutein + zeaxanthin (μg)	91
Peonidin 3-O-galactoside (mg)	22.02
Peonidin-3-O-arabinoside (mg)	9.61
Peonidin 3-O-glucoside (mg)	4.16
Cyanidin 3-O-galactoside (mg)	8.89
Cyanidin 3-O-arabinoside (mg)	4.47
Cyanidin 3-O-glucoside (mg)	0.74
Quercetin 3-O-galactoside (mg)	10.81
Quercetin 3-O-arabinoside (mg)	4.94
Quercetin 3-O-rhamnoside (mg)	6.17
Myricetin 3-O-arabinoside (mg)	5.30
Kaemperfol 3-O-glucoside (mg)	0.87
Benzoic acid (mg)	48.1
2,4-Didydroxybenzoic acid (mg)	0.8
3-Hydroxybenzoic acid (mg)	0.41
4-Hydroxybenzoic acid (mg)	0.42
Vanillic acid (mg)	0.69
Caffeic acid (mg)	0.38
Cinnamic acid (mg)	0.16
Ferulic acid (mg)	0.81
PACs (mg)	217.6
Phloridzin (mg)	12
Ursolic acid (mg)	60–100

[a] Values are obtained from https://fdc.nal.usda.gov/fdc-app.html#/food-details/171722/nutrients, http://phenol-explorer.eu/contents/food/74, Gu et al. (2004), Turner et al. (2007), Neto and Vinson (2011).

their contents have been employed for the standardization of cranberry products. PACs are generally measured using HPLC and 4-(dimethylamino) cinnamaldehyde (DMAC) colorimetric assays (Birmingham et al., 2021,

© Burleigh Dodds Science Publishing Limited, 2022. All rights reserved.

Gao et al., 2018). While HPLC can quantify individual PACs with different degrees of polymerization, DMAC colorimetric assays measure the overall PAC content and are more commonly used to standardize cranberry products, particularly nutritional supplements. Standardizing cranberry products with a particular cranberry constituent such as PACs is crucial because the quality of cranberry products can be tainted by adulterations with other ingredients, such as anthocyanin-rich ingredients, with a similar color.

Cranberries also contain phloridzin, a dihydrochalcone (12 mg/100 g), and ursolic acid, a triterpenoid (60–100 mg/100 g), and insignificant amounts of lutein (Turner et al., 2007, Neto and Vinson, 2011). Soluble xyloglucan and pectic oligosaccharides, which are non-digestible carbohydrates, are the most recently recognized bioactives in cranberries, particularly in cranberry pomace, a by-waste of juice production (Coleman and Ferreira, 2020). These oligosaccharides are present at ~20% *w/w* in many cranberry materials, especially dehydrated cranberry pomace powder, and display anti-adhesion activity against bacteria (Sun et al., 2019).

3 Health benefits

3.1 Urinary tract infections

Urinary tract infections (UTIs) affect 150 million people each year worldwide. In the United States alone, approximately 10.5 million office visits (Schappert and Rechtsteiner, 2011) and 1.5 million emergency department visits (Niska et al., 2010) occur every year due to UTI symptoms. Clinically, UTIs are categorized as uncomplicated or complicated, with the latter defined as those associated with factors that compromise the urinary tract or host defense, including urinary obstruction, urinary retention caused by neurological disease, immunosuppression, renal failure, renal transplantation, and the presence of foreign bodies such as calculi, indwelling catheters, or other drainage devices, as well as UTIs that occur in women during pregnancy and men (Gupta et al., 2011, Nicolle and AMMI Canada Guidelines Committee, 2005, Dason et al., 2011). Uncomplicated UTIs typically affect individuals who are otherwise healthy women without structural or neurological urinary tract abnormalities. The self-reported annual incidence of uncomplicated UTI in women is 12%, and by the age of 32 years, half of all women report having had at least one UTI (Foxman and Brown, 2003). Additionally, the rate of recurrence in the 12 months following an initial uncomplicated UTI in women is high, ranging from 25% to 44% (Mabeck, 1972, Foxman, 1990, Ikäheimo et al., 1996). The high incidence and recurrence rate of uncomplicated UTI among women, along with the rapid rise of multi-drug resistant uropathogens, provide support for the timely assessment of cranberries for decreased risk of uncomplicated recurrent UTI among women.

© Burleigh Dodds Science Publishing Limited, 2022. All rights reserved.

In 2020, the US FDA approved a qualified health claim that consuming one serving (8 oz) each day of a cranberry juice beverage or 500 mg each day of cranberry dietary supplement may help reduce the risk of recurrent UTI in healthy women (FDA, 2020). In approving the qualified health claim, the FDA relied on the evidence provided by seven publications reporting on eight intervention studies (five intervention studies on cranberry juice beverages, i.e. Barbosa-Cesnik et al., 2011, Maki et al., 2016, Stapleton et al., 2012, Stothers, 2002, Takahashi et al., 2013; and three intervention studies on cranberry dietary supplement, i.e. Stothers, 2002, Vostalova et al., 2015, Walker et al., 1997). Of the five studies on cranberry juice, two (Maki et al., 2016, Stothers, 2002) demonstrated a statistically significant benefit and one demonstrated mixed results whereby no statistically significant beneficial effect was observed when subjects with uncomplicated UTI (n = 170) consumed a daily dose of 125 mL of a cranberry juice beverage (65% of cranberry juice), but a statistically significant beneficial effect was observed in a sub-analysis of women aged 50 years and older (n = 118). The remaining two studies (Barbosa-Cesnik et al., 2011, Stapleton et al., 2012) showed no effect of consuming a cranberry juice beverage on risk reduction of recurrent UTI. With regard to cranberry supplementation, the three identified studies (Stothers, 2002, Vostalova et al., 2015, Walker et al., 1997) all demonstrated a statistically significant benefit of a cranberry dietary supplement consumption on the risk of recurrent UTI in healthy women with a history of UTI. Because of the limited number of randomized controlled studies (i.e. three on cranberry supplementation and five on cranberry juice with high and moderate methodological quality but with inconsistent results), FDA has concluded that the current evidence provides only qualified support for the scientific validity of the relationship between cranberry juice/supplements and recurrent UTIs.

The specificity of the qualified health in recurrent UTIs is consistent with evidence suggesting a difference in the efficacy of cranberries against uncomplicated vs. complicated UTIs. According to a meta-analysis by Wang et al. (2012), cranberry-containing products were effective in women with recurrent UTIs, but not in populations with potentially complicated UTIs such as those with neuropathic bladder, elderly patients, and pregnant patients. A Cochrane meta-analysis published in the same year as the Wang et al. (2012) meta-analysis concluded that cranberries are not effective against UTIs (Jepson et al., 2012). A review by Liska et al. (2016) suggested that one of the reasons for the difference between the Cochrane and Wang et al. (2012) meta-analysis conclusions is that the overall conclusion made by Jepson et al. (2012) was more heavily influenced by results from studies in populations with complicated UTIs whereas the Wang et al. (2012) meta-analysis weighted the evidence relatively equally across the populations. A more recent meta-analysis by Fu et al. (2017) focused only on uncomplicated UTIs (recurrent

© Burleigh Dodds Science Publishing Limited, 2022. All rights reserved.

UTIs in healthy women) and concluded that cranberry reduced the risk of UTIs. Additionally, another recent meta-analysis (Raguzzini et al., 2020) showed that cranberries are ineffective against UTIs in patients with spinal cord injury when compared with control. Although the mechanism by which cranberries affect UTI is still under investigation, the differential effects of cranberry products on uncomplicated and complicated UTI may lie in their pathogenesis and the anti-adhesion activities of cranberry constituents. For example, uropathogenic *E. coli* is the most common causative pathogen of uncomplicated UTI, accounting for ~80% of infections but only ~21% for complicated UTI (Stamm and Hooton, 1993, Hidron et al., 2008).

3.2 H. pylori-*induced gastric infection*

Helicobacter pylori (*H. pylori*), a gram-negative bacterium pathogen, was estimated to affect more than half of the world's population and is associated with chronic gastritis, peptic ulcer disease, mucosa-associated lymphoid tissue (MALT) lymphoma, and gastric cancer (Guevara and Cogdill, 2020). *H. pylori* infection begins with the adhesion of the bacteria to the gastric epithelium, followed by inflammation and ulcer (Wang et al., 2014). Standard regimens of *H. pylori* eradication include an array of antibiotic treatments, but their substantially reduced effectiveness due to antibiotic resistance has raised a need for alternative treatments (Savoldi et al., 2018). Since the adhesion of *H. pylori* to the gastric epithelium is the initial step of the infection, cranberry constituents are anticipated to be capable of diminishing *H. pylori* infection via their anti-bacterial adhesion activity. The potential benefit of cranberries in *H. pylori* infection was first reported in a prospective, randomized, double-blind, placebo-controlled trial with 189 adults with *H. pylori* infection in Shandong, China (Zhang et al., 2005). This trial showed that daily consumption of 500 mL cranberry juice for 90 days resulted in *H. pylori* eradication, assessed using a 13C-urea breath test, in 14% of participants as compared to 5% consuming placebo ($P<0.05$). While this result is significant, it appears the positive response rate is low. Additionally, such effect was only observed in women, but not in men. The comparable eradication efficacy of cranberry was also reported in children (Gotteland et al., 2008).

Because there is no need for absorption and transportation of specific active constituents to the target tissue, theoretically, any cranberry constituents can play a role in the eradication of *H. pylori*. A recent PAC dose-response trial with 522 *H. pylori*-positive adults in Shandong, China, was designed to examine whether PACs would be the main contributor (Li et al., 2020). The results of this double-blind, randomized, placebo-controlled trial showed that consumption of cranberry juice providing 88 mg for 8 weeks led to a 20% *H. pylori*-negative rate as compared to 7.6, 4.5, and 7.3% following 44, 23, and 0 mg/d (placebo)

© Burleigh Dodds Science Publishing Limited, 2022. All rights reserved.

PAC, respectively. Since other non-PAC polyphenols were not controlled, whether PACs are the primary effector remains obscure. Interestingly, the null effect was observed when cranberry polyphenols were consumed in an encapsulated powder form (Li et al., 2020), suggesting the effectiveness of cranberry constituents against *H. pylori* depends on formulation forms.

As antibiotic treatments are standard for *H. pylori* infection, cranberry constituents may work with antibiotics in a synergistic/additive manner. This concept was suggested in a meta-analysis showing that the addition of antioxidants (vitamin C or E, N-acetylcysteine, curcumin, cranberry) to amoxicillin-clarithromycin-based therapy enhanced the eradication rate from 68.6% to 81.3% (Yang-Ou et al., 2018). Shmuely et al. (Shmuely et al., 2007) also noted that a combination of triple therapy (omeprazole, amoxicillin, and clarithromycin) with daily consumption of 500 mL cranberry juice significantly increased the *H. pylori* eradication rate in Israeli female patients from 86% with triple therapy alone to 95%. Similarly, an improved eradication rate (74% vs. 89%) was reported in another trial with Iranian patients with peptic ulcer disease receiving lansoprazole, clarithromycin, and amoxicillin and 500 mg/d cranberry capsules (Seyyedmajidi et al., 2016). Cranberry alone or together with standard antibiotic therapies at adequate PAC or polyphenol contents has the potential for suppressing *H. pylori* infection in some responders. Future studies are warranted to characterize factors for the responder effect and identify the actual compounds responsible for the *H. pylori* eradication.

3.3 Dental health

Oral health is essential to supporting nutrition needs and thus is an integral element of overall health and well-being (Peres et al., 2019). Oral diseases are chronic and progressive in nature and include dental caries (tooth decay), periodontal (gum) disease, and oral cancers. Dental caries is the most prevalent disease in the world and is caused by acid by-products produced from bacterial fermentation of free sugars in the dental plaque biofilm composing a complex microbial community (Peres et al., 2019). Thus, bacterial populations residing in the human mouth have a marked impact on oral health as colonization and proliferation of pathogenic ones can increase the risk of dental caries and periodontal disease (Marsh, 2018, Seneviratne et al., 2011). Given that cranberry constituents, particularly polyphenols, display anti-inflammatory, anti-microbial, and anti-bacteria adhesion, cranberry products are anticipated to help maintain/improve oral health (Philip and Walsh, 2019).

Most evidence on the beneficial effects of cranberry constituents on dental caries and periodontal disease were gathered from *in vitro* experiments (Greene et al., 2020, Feghali et al., 2012, Sánchez et al., 2020, de Medeiros et al., 2016). For example, Philip et al. (2019) demonstrated in an *in vitro* experiment

© Burleigh Dodds Science Publishing Limited, 2022. All rights reserved.

that cranberry extracts reduced saliva-derived polymicrobial biofilm biomass, acidogenicity, exopolysaccharide (EPS)/microbial biovolumes, bacterial counts, and the relative abundance of specific caries- (*Streptococcus sobrinus* and *Provotella denticola*) and health-associated bacteria (*Streptococcus sanguinis*).

Clinical evidence supporting the potential of cranberry products for oral health remains scarce. A recent human trial tested the effect of adding cranberry polyphenol-rich extract to a casein phosphopeptide-amorphous calcium phosphate (CPP-ACP) dentifrice changed the ecology of oral microbiome in a parallel, 3-group, double-blind, randomized, and controlled trial with 90 participants with at least 4 fully erupted permanent maxillary teeth (Philip et al., 2020). The results showed that cranberry polyphenols were more effective in decreasing the bacterial loads of two caries-associated bacterial species (*Streptococcus mutans* and *Veillonella parvula*) and increasing two health-associated bacterial species (*Neisseria flavescens* and *Streptococcus sanguinis*) as compared to CPP-ACP and fluoride control dentifrices, suggesting the possibility of formulating polyphenol-rich cranberry extracts in oral care products to support healthy oral microbial communities. An earlier parallel randomized trial with 50 patients with gingivitis showed that as compared to water, daily consumption of 750 mL cranberry functional beverage consisting of cranberry juice (20% by volume), apple juice (80%), and ground cinnamon (0.25 g/100 mL) inhibited dental plaque deposition and gingival inflammation but did not affect the count of *Streptococcus mutans* in dental plaques (Woźniewicz et al., 2018).

3.4 Gut health via microbiota

Thousands of microbes are residing in/on the human body, particularly in the colon where they interact intimately with foods and human host. Although a healthy microbiome profile has not been established, the composition and metabolic capacity of the gut microbiota appear critical to human health locally and systematically, such as the case for obesity and insulin resistance, type 2 diabetes mellitus, gastrointestinal disorders, and recurrent infection (Fassarella et al., 2021). Microbial diversity, metabolic flexibility, microbe-microbe and host-microbe interactions are subject to perturbations by fecal microbiota transplantation, supplementation with probiotics or non-digestible carbohydrates, and dietary modifications (Fassarella et al., 2021). Given that cranberry constituents decrease UTIs via their bactericidal and anti-bacteria adhesion activities, their impact on the gut microbiome is anticipated (Zhao et al., 2020).

Several human trials have been conducted to evaluate how cranberry modulates the gut microbiome. A very recent trial showed that daily consumption of a cranberry beverage for 24 weeks did not affect the overall

© Burleigh Dodds Science Publishing Limited, 2022. All rights reserved.

microbiome composition, microbial diversity, functional pathways, and relative abundances of most bacterial taxa in women with UTI but decreased one unnamed *Flavonifractor* species (OTU41) (Straub et al., 2021). This bacteria is associated with negative human health and relates to the transport and metabolism of tryptophan and cobalamin. The effect of cranberry constituents on the gut microbiome was also observed in a small-scale randomized, crossover, controlled-feeding trial with 11 healthy adults (Rodríguez-Morató et al., 2018). As compared to the animal-based control diet, consumption of cranberry diet consisting of 30 g/day freeze-dried whole cranberry powder for 5 days modified 9 taxonomic clades, including a decrease in the abundance of *Firmicutes* and increase in *Bacteroidetes*, *Lachnospira,* and *Anaerostipes* and attenuated the increase in carcinogenesis-related secondary bile acids and decrease in acetic and butyric acids caused by the animal-based control diet. Similarly, another small-scale trial showed that 2-week consumption of sweetened dried cranberries (42 g/d) with lunch tended to change the relative abundance of several bacterial taxa, particularly with an increase in *Akkermansia* bacteria and a decrease in the ratio of *Firmicutes* to *Bacteroidetes* (Bekiares et al., 2018). The main difference in study products (juice vs. whole cranberry) and dose may contribute to the discrepancy between the two latter trials (Rodríguez-Morató et al., 2018, Bekiares et al., 2018) and that of the first more robust study (Straub et al., 2021). Ursolic acid, undetectable in commercial cranberry juice, may also contribute to the gut microbiome modulating effect of whole cranberry products used in the two latter trials (Rodríguez-Morató et al., 2018, Bekiares et al., 2018) as suggested by the decrease in the ratio of *Firmicutes* to *Bacteroidetes* and increase in the growth of short-chain fatty acid (SCFA)-producing bacteria in preclinical studies (Neto, 2007, Hao et al., 2020). Additionally, non-digestible complex cranberry carbohydrates, such as xyloglucans, can be utilized by microbes in the large intestine to generate products that can influence the gut microbiome via a cross-feeding interaction between microbes (Özcan et al., 2017).

3.5 Cardiometabolic diseases

Cardiometabolic diseases are a cluster of diseases such as diabetes, hypertensive and ischemic heart diseases, and metabolic dysfunction-associated fatty liver disease that share a number of common risk factors, including obesity, prehypertension, pre-diabetics, hyperlipidemia, and insulin resistance. Drug therapies are frequently and effectively applied, although their undesirable side effects support and promote lifestyle modifications, such as diet and exercise, for the management of these risk factors. Given that cranberry products contain antioxidant phytochemicals, their potential for the amelioration of these risk factors has been examined in preclinical and clinical studies. A

© Burleigh Dodds Science Publishing Limited, 2022. All rights reserved.

recent meta-analysis of 12 randomized trials comprising 496 participants showed that cranberry products significantly improved systolic blood pressure (−3.63 mmHg; 95% CI: −6.27, −0.98) and body mass index (−0.30 kg/m^2, 95% CI: −0.57 to −0.02) but did not affect lipid profile, blood glucose and insulin, diastolic blood pressure, waist circumference, C-reactive protein, and intercellular adhesion molecule (Pourmasoumi et al., 2020). Interestingly, the efficacy of cranberry products is age-dependent such that systolic pressure improved in people>50 years and HDL improved in people<50 years. This divergence leads to a notion that personalized regimens for the prevention or management of cardiometabolic diseases must be made based on individual's metabolic, genetic, and physiological conditions. Additionally, the impact of product type (juice vs. supplement), bioactive composition of intervention products, subject demographics, dose, and intervention on the efficacy of cranberry products on risk factors of cardiometabolic diseases must be elucidated. For example, Rodriguez-Mateos et al. (2016) identified 12 plasma polyphenol metabolites that significantly correlated with enhanced endothelial function, as measured using a flow-mediated dilation test, including ferulic and caffeic acid sulfates, quercetin-3-O-ß-D-glucuronide and a ⍰-valerolactone sulfate, suggesting bioactive compounds may not be present in cranberry products but derived from bacterial metabolism of cranberry polyphenols.

Metabolic dysfunction-associated fatty liver disease shares many risk factors with other cardiometabolic diseases, such as obesity and insulin resistance (Eslam et al., 2020). Polyphenol-rich fruits, spices, teas, coffee, and other plant foods are helpful for fatty liver disease by ameliorating hepatic steatosis, oxidative stress, inflammation, and apoptosis (Li et al., 2021). A recent double-blind placebo-controlled randomized trial with 41 patients with the non-alcoholic fatty liver disease showed a larger improvement in alanine aminotransferase (index of liver damage and inflammation) and insulin resistance following 12-week supplementation of a cranberry supplement (288 mg extract/day, equivalent to 26 g dried cranberry) compared to placebo (Hormoznejad et al., 2020).

3.6 Cognition

Dietary supplementations with fruit or vegetable extracts ameliorate age-related declines in learning, memory, motor performance, and neuronal signal transduction in a rat model (Shukitt-Hale et al., 2005). Additionally, the potential cognitive benefits of polyphenols consumed in the form of an extract or whole food have been supported by preclinical and observational studies and some clinical studies (Fraga et al., 2019, Gildawie et al., 2018). Increased intake of flavonoids and anthocyanins, particularly from berries, were strongly associated with a slower rate of cognitive decline in 16010 older women (Devore et al.,

© Burleigh Dodds Science Publishing Limited, 2022. All rights reserved.

2012). Another observational study also showed intake of anthocyanins, flavones, flavonols, and phenolic acids was associated with enhanced performance in verbal and episodic memory tasks 13 years later but was negatively associated with executive functioning in 2574 middle-aged adults (Kesse-Guyot et al., 2012). Several studies have reported positive clinical evidence with selected flavonoid subfamilies and foods, such as cocoa flavanols and anthocyanin-rich foods (Mastroiacovo et al., 2015, Sloan et al., 2021, Whyte et al., 2018, Travica et al., 2020, Kent et al., 2017). A recent study showed that 16-week supplementation of 160 mg/day of purified anthocyanins derived from bilberry and blackcurrant improved memory and executive function in older adults with mild cognitive impairment (Bergland et al., 2019). Natural PACs are reported to attenuate pathological features of Alzheimer's disease, including extracellular amyloid deposits and neurofibrillary tangles (Zhao et al., 2019). The exact mechanisms by which polyphenols modulate cognitive functions remain to be elucidated, and emerging mechanisms include increasing cerebral blood flow (CBF), accelerating brain oxygenation, improving insulin sensitivity, and modulating brain activity (Fraga et al., 2019, Gratton et al., 2020).

As cranberry products contain flavanol, anthocyanins, and other polyphenols, their consumption is anticipated to be beneficial to cognition, but preclinical and clinical evidence is scarce. In an older trial with 50 cognitively intact older adults, as compared to placebo, consumption of 32 ounces/day of a beverage containing 27% cranberry juice for 6 weeks did not affect thinking processes and mood even though the ability to remember tended to improve (Crews et al., 2005). A positive effect of dried water extract of frozen cranberries (2% of diet) was noted in an aged rat experiment (Shukitt-Hale et al., 2005). Additionally, water-soluble compounds in cranberries, including some polyphenols, enhanced neuronal signal transduction as measured by striatal dopamine release and ameliorated deficits in motor performance (muscle tone and strength and balance) and hippocampal HSP70 neuroprotection. Thus, the effect of dietary polyphenols, including those in cranberries, on cognitive functions is promising, but more clinical evidence is needed to substantiate the benefits via diet, supplementation, or both.

3.7 Skin health

Human skin is not only a physical barrier against biological, chemical, and physical insults but also contributes to water and electrolyte balance, thermoregulation, and immunity (Michalak et al., 2021, Feng et al., 2021). Skin health is subject to the influence of natural aging and photoaging, with the latter mainly caused by free radicals derived from environmental factors, including UV radiation, pollution, and cigarette smoke (Michalak et al., 2021). Among them, the harm of UV radiation is well appreciated as it decreases the

© Burleigh Dodds Science Publishing Limited, 2022. All rights reserved.

synthesis of collagen fibrin and elastic fibrin, augments abnormal elastic fibrin, enables disappearance of the extracellular matrix, and causes protein and DNA damages (Feng et al., 2021). Thus, nutrients that support or enhance antioxidant defense are anticipated to promote skin health. For example, carotenoids (e.g. lutein) and polyphenols (e.g. apigenin, quercetin, curcumin, silymarin, and PACs), and antioxidant vitamins and minerals are appreciated for their potential benefits in protection against photoaging (Dini and Laneri, 2019, Hernandez et al., 2020). Additionally, a balanced diet with an adequate intake of proteins, carbohydrates, and fats is essential to skin health because the skin is a tissue with a high turnover rate and requires ample nutrients for cellular generation (Cao et al., 2020).

While clinical evidence of cranberry products on skin health remains lacking, their potential is partially supported by positive results of other polyphenol-rich foods or nutraceuticals (Pérez-Sánchez et al., 2018). Several human trials had shown that Pycnogenol®, a standardized French maritime pine bark extract rich in catechins and B-type procyanidins and phenolic acids (caffeic, ferulic, and p-hydroxybenzoic acids), was protective against UV light-induced skin damage and promoted de novo skin collagen (Saliou et al., 2001, Marini et al., 2012, Furumura et al., 2012). A supplement containing citrus and rosemary extracts, NutroxSun®, enriched in naringenin, phenolics, and diterpenes (carnosic acid) was also noted to increase UV radiation-induced minimal erythema dose (a measure of skin photosensitivity) by 56% and decrease wrinkle depth and increase skin elasticity (Pérez-Sánchez et al., 2014, Nobile et al., 2016). Whether cranberry polyphenols and other constituents possess properties toward skin health is uncertain at this time, a clinical trial showed that oral intake of cranberry proanthocyanidin and antioxidative vitamins A, C, and E significantly decreased the degree of pigmentation (epidermal melisma) in the left malar and right malar regions of 60 middle-aged Filipino (Handog et al., 2009), suggesting a potential of cranberry products for treating natural aging and photoaging-related skin pigmentation.

3.8 Cancer

Cranberry and its components, particularly flavonols, anthocyanins, and PCAs, display chemopreventive properties by reducing oxidative stress and inflammation, inhibiting cell proliferation and angiogenesis, inducing cell apoptosis, and attenuating metastasis (Mantzorou et al., 2019, Zhao et al., 2020). The putative benefits of cranberry are demonstrated in preclinical studies. Zhao et al. (2020) recently summarized that cranberry juice extract and PCA and flavonoid fractions diminished the cell viability of 41 cell lines, which included 16 cancer types. However, the utility of the *in vitro* data is questionable, such as in lung cancer (Kresty et al., 2011), because most cranberry PCAs are not

© Burleigh Dodds Science Publishing Limited, 2022. All rights reserved.

absorbable (Walsh et al., 2016). Nevertheless, the anti-cancer potential of cranberry constituents for gastrointestinal cancers warrants further examination as bioactives in cranberry can exert actions without the initial absorption process. A preclinical study showed that triterpenes, sterols, and polyphenol-rich fractions were effective in suppressing tumor metrics in mice with colitis-associated colon cancer (Wu et al., 2020). Additionally, the efficacy of cranberry polyphenols against colon cancer may be amplified by combining with cell wall biomass of Lactobacillus probiotics (Desrouillères et al., 2020), supporting the development of products containing both probiotics and cranberry polyphenols. Moreover, the potential additive/synergistic effects of different cranberry constituents warrant exploration. For example, Xiao et al. (2015) noted that dried cranberries were more protective than a polyphenol-rich cranberry extract against dextran sodium sulfate-induced colitis symptoms in mice. While preclinical evidence supports the promising anti-cancer potentials of cranberry and its components, clinical and observational data remain scarce. Student et al. (2016) reported in a randomized, parallel trial that 30 days of 1500 mg cranberry fruit powder supplementation decreased serum prostate-specific antigen in patients with prostate cancer but did not affect cancer staging scores evaluated after prostatectomy, as compared to placebo. In an earlier study, the same group also observed a reduction in serum prostate-specific antigen (PSA) in men at risk of prostate disease and with elevated PSA and clinically confirmed chronic non-bacterial prostatitis (Vidlar et al., 2010).

4 Future directions

Cranberry products are protective against recurrent UTIs, but the exact mechanism(s) of action has not been elucidated (González de Llano et al., 2020, Scharf et al., 2020). The observed benefits have been postulated to be due to A-type PACs in cranberry via their anti-adhesive activity against uropathogenic *E. coli* by their interaction with [D-Gal-(1-4)-⍰-D-Gal]-binding P-fimbriae (Scharf et al., 2020). However, PACs are poorly bioavailable and do not accumulate to amounts with detectable anti-adhesive efficacy *in vitro* (de Llano et al., 2015, Walsh et al., 2016), suggesting that other mechanisms may be responsible. Additionally, other constituents in whole cranberries may work additively or synergistically with polyphenols and others to support health. The result of studies on the anti-adhesive activity of PAC-free cranberry extracts supports the presence of other bioactive constituents, such as soluble xyloglucan and pectic oligosaccharides (Sun et al., 2019, Rafsanjany et al., 2015, González de Llano et al., 2020, Coleman et al., 2019). Xyloglucan and pectic oligosaccharides are present at ~20% *w/w* in many cranberry materials, but their effects on UTI and other health conditions remained to be examined in human trials. Furthermore, the potential contribution of cranberry-derived changes in composition and

© Burleigh Dodds Science Publishing Limited, 2022. All rights reserved.

metabolic capacity of the gut microbiota in the prevention of UTI recurrence needs to be tested because the intestine is a possible source of uropathogenic bacteria. Concrete evidence on the exact mechanisms of action is expected to inform the development of cranberry products formulated with the most efficacious components against UTI.

Bioefficacy of cranberry constituents may be enhanced when they are formulated with other ingredients that can generate synergistic or additive interactions. Combining cranberry extracts and *Lactobacillus* probiotic decreased the number of women who experienced recurrent UTIs than placebo in a 26-week trial with pre-menopausal adult women (Koradia et al., 2019). Along the same lines, the *H. pylori* eradication rate of cranberry products can be improved by combining with other alternative treatments, such as omega-3 fatty acids (Zare Javid et al., 2020). However, these two studies were not designed to examine whether there was a synergistic or additive interaction between cranberry and other functional ingredients, although it should be noted that a synergy between *Lactobacillus acidophilus* and cranberry c-PAC in reduction in invasiveness of pathogenic *E. coli* was noted *in vitro* (Polewski et al., 2016). Regardless, the development of cranberry products containing other functional ingredients, such as probiotics and others, for all health conditions should be pursued further (González de Llano et al., 2020).

More research is warranted to establish the causal evidence of cranberry products in cardiometabolic diseases, cognitive functions, skin health, and chemoprevention (Grammatikopoulou et al., 2020, Zhao et al., 2020). Additionally, cranberry constituents, particularly polyphenols, display enormous potential for supporting oral health, but there is a need for more clinical evidence, as well as the development of innovative oral health products containing cranberry constituents such as cranberry polyphenols with probiotics. Moreover, the benefits of novel products, such as cranberry seed oil containing flavonoids, salicylic acid, omega-3 fatty acids, for cardiometabolic health must be substantiated in adequately designed human trials. Clinical investigation for the development of cranberry products to alleviate metabolic dysfunction-associated fatty liver disease is also encouraged because of the increasing prevalence of the disease in developed and developing countries.

In the personalized nutrition era, the effectiveness of cranberry products in health is individual-dependent, likely due to variability in the absorption and metabolism of active constituents in the small intestine, microbial metabolism in the large intestine, background diet, host genetics, and other physiological factors. For example, phenyl-�-valerolactones, proven *in vitro* to have anti-adhesive activity against uropathogenic *E. coli* in bladder epithelial cells (Mena et al., 2017), may contribute to UTI protection but their production can vary between individuals due to variation in the gut microbiota composition (Feliciano et al., 2017). Understanding the magnitude of the individual variability

© Burleigh Dodds Science Publishing Limited, 2022. All rights reserved.

will help inform advancement in cranberry product developments for certain health conditions and specific populations.

The bioefficacy of cranberry products depends mainly on the bioavailability of bioactive constituents and/or their metabolites in target tissues. For example, the protection against uncomplicated UTI is anticipated to be maximized if cranberry A-type PACs can reach the urinary tract at higher concentrations by using novel formulation technologies, such as encapsulation delivery systems (Delfanian and Sahari, 2020). Thus, research on technologies that enhance the bioavailability of bioactive constituents in target tissues is warranted.

Products derived from foods are generally considered safe but are not totally without risk. Untoward effects of some specific phytochemicals consumed at a high dose have been appreciated, and the adversities can be escalated particularly when there are interactions with drugs (Ronis et al., 2018). Cranberry products, including supplements, are safe for moderate consumption by most people. However, there are some mixed reports in the literature showing an interaction between cranberry and warfarin, in which significant bleeding can occur (Tan and Lee, 2021, Srinivas, 2013), possibly through the modulation of cytochrome P450 isozymes responsible for warfarin metabolism (Suvarna et al., 2003). Interestingly, warfarin pharmacokinetics appear not to be affected by consumption of cranberry juice but altered by cranberry supplements, suggesting a need for caution when using concentrated cranberry products. When consuming concentrated cranberry products, the potential drug interaction may also be dependent on the length of consumption, with deleterious effects observed >3 weeks (Srinivas, 2013). While more clinical evidence is warranted to substantiate the cranberry-warfarin interaction in susceptible populations, it is prudent for warfarin users not to substantially increase cranberry intake, particularly products containing cranberry phytochemical extracts.

5 Conclusion

American cranberries are a rich source of phytochemicals, particularly A-type PAC and anthocyanins. Cranberries and cranberry-derived products are commonly consumed in the American diet, and their consumption is best known for protection against the recurrence of UTIs. More significantly, this health benefit has received a US FDA qualified health claim in 2020. The health benefits of cranberries can also be extended to bacteria-related health conditions, such as dental and gastrointestinal health, and others, such as cardiometabolic and skin health, cognition, and cancer (Table 2). Among these health benefits, reduction in blood pressure and body weight is supported by the results of a meta-analysis (Pourmasoumi et al., 2020). Other health benefits require more clinical evidence to realize the potential of cranberry

© Burleigh Dodds Science Publishing Limited, 2022. All rights reserved.

Table 2 Summary of health benefits of cranberries

Health benefits	Mechanism of actions	Evidence[a]	Level of evidence[b]
Uncomplicated UTI	• Anti-E. coli adhesion activities	• Clinic • Positive	Convincing
Complicated UTI		• Clinic • Null	Convincing
***H. pylori* infection**	• Anti-bacterial adhesion activity	• Clinic • Positive	Probable
Dental health	• Anti-inflammatory activity • Anti-microbial activity • Anti-bacterial adhesion activity	• Clinic • In vitro • Positive	Possible
Gut health	• Bactericidal activity • Anti-bacterial adhesion activity	• Clinic • Positive	Possible
Cardiometabolic diseases	• Antioxidant activity • Modulating insulin sensitivity	• Clinic • Positive (blood pressure and body weight)	Convincing
Cognition	• Increasing cerebral blood flow • Accelerating brain oxygenation • Improving insulin sensitivity • Modulating brain activity	• Pre-clinic • Positive	Insufficient
Skin health	• Antioxidant activity	• Clinic • Positive	Possible
Cancer	• Antioxidant activity • Anti-inflammatory activity • Inhibiting cell proliferation • Inhibiting angiogenesis • Inducing cell apoptosis • Attenuating metastasis	• Pre-clinic • Positive	Insufficient

Abbreviation: UTI, urinary tract infection
[a] Clinic: human data; pre-clinic: animal and/or cell culture data.
[b] Convincing evidence is designated with support of systematic review and meta-analysis; probably evidence support of several human studies; possible support of a few human studies; insufficient support of preclinical studies.

© Burleigh Dodds Science Publishing Limited, 2022. All rights reserved.

products. Furthermore, more research needs to be conducted to elucidate the mechanisms of actions by which cranberry constituents support health benefits, for example, whether other cranberry constituents, such as xyloglucans, besides A-type PAC contribute to the protection of UTI recurrence and modulation of gut microbiota. This information is anticipated to inform the development of cranberry products formulated with the most efficacious components. Efforts on establishing additive/synergistic interaction of cranberry components with other ingredients are encouraged, such as antibiotics and probiotics for bacteria-related health issues. Finally, the development of novel technologies that enhance the bioavailability of bioactive constituents of cranberries in target tissues is warranted to maximize bioefficacy.

6 Where to look for further information

New information about the nutraceutical properties of cranberries is anticipated to be added to the literature and can be found in the PubMed database (https://pubmed.ncbi.nlm.nih.gov/), using cranberry and nutraceuticals/functional foods as keywords for the search.

7 References

Barbosa-Cesnik, C., Brown, M. B., Buxton, M., Zhang, L., Debusscher, J. and Foxman, B. 2011. Cranberry juice fails to prevent recurrent urinary tract infection: Results from a randomized placebo-controlled trial. *Clin. Infect. Dis.* 52(1), 23–30.

Bekiares, N., Krueger, C. G., Meudt, J. J., Shanmuganayagam, D. and Reed, J. D. 2018. Effect of sweetened dried cranberry consumption on urinary proteome and fecal microbiome in healthy human subjects. *Omics* 22(2), 145–153.

Bergland, A. K., Soennesyn, H., Dalen, I., Rodriguez-Mateos, A., Berge, R. K., Giil, L. M., Rajendran, L., Siow, R., Tassotti, M., Larsen, A. I. and Aarsland, D. 2019. Effects of anthocyanin supplementation on serum lipids, glucose, markers of inflammation and cognition in adults with increased risk of dementia – A pilot study. *Front. Genet.* 10, 536.

Bhagwat, S. and Haytowitz, D. B. 2015. USDA database for the proanthocyanidin content of selected foods, release. Available at: https://data.nal.usda.gov/dataset/usda-database-proanthocyanidin-content-selected-foods-release-2-2015 [Accessed April 6, 2021]. Nutrient Data Laboratory, Beltsville Human Nutrition Research Center, Anaesthetics Research Society, United States Department of Agriculture (vol. 2).

Bhagwat, S. and Haytowitz, D. B. 2016. USDA database for the flavonoid content of selected foods. Release 3.2 (November 2015). Available at: https://data.nal.usda.gov/dataset/usda-database-flavonoid-content-selected-foods-release-32-november-2015 [Accessed April 6, 2021]. Nutrient Data Laboratory, Beltsville Human Nutrition Research Center, Anaesthetics Research Society, United States Department of Agriculture.

© Burleigh Dodds Science Publishing Limited, 2022. All rights reserved.

Birmingham, A. D., Esquivel-Alvarado, D., Maranan, M., Krueger, C. G. and Reed, J. D. 2021. Inter-laboratory validation of 4-(dimethylamino) cinnamaldehyde (DMAC) assay using cranberry proanthocyanidin standard for quantification of soluble proanthocyanidins in cranberry foods and dietary supplements, first action official MethodSM: 2019.06. *J. AOAC Int.* 104(1), 216–222.

Cao, C., Xiao, Z., Wu, Y. and Ge, C. 2020. Diet and skin aging - From the perspective of food nutrition. *Nutrients* 12(3), 870.

Central, F. 2020. Cranberries, raw. Available at: https://fdc.nal.usda.gov/fdc-app.html#/food-details/1102706/attributes [Accessed April 6, 2021]. Agricultural Research Service, United States Department of Agriculture.

Coleman, C. M., Auker, K. M., Killday, K. B., Azadi, P., Black, I. and Ferreira, D. 2019. Arabinoxyloglucan oligosaccharides may contribute to the antiadhesive properties of porcine urine after cranberry consumption. *J. Nat. Prod.* 82(3), 589–605.

Coleman, C. M. and Ferreira, D. 2020. Oligosaccharides and complex carbohydrates: A new paradigm for cranberry bioactivity. *Molecules* 25(4), 881.

Crews, W. D., JR, Harrison, D. W., Griffin, M. L., Addison, K., Yount, A. M., Giovenco, M. A. and Hazell, J. 2005. A double-blinded, placebo-controlled, randomized trial of the neuropsychologic efficacy of cranberry juice in a sample of cognitively intact older adults: Pilot study findings. *J. Altern. Complement. Med.* 11(2), 305–309.

Dason, S., Dason, J. T. and Kapoor, A. 2011. Guidelines for the diagnosis and management of recurrent urinary tract infection in women. *Can. Urol. Assoc. J.* 5(5), 316–322.

de Llano, D. G., Esteban-Fernández, A., Sánchez-Patán, F., Martínlvarez, P. J., Moreno-Arribas, M. V. and Bartolomé, B. 2015. Anti-adhesive activity of cranberry phenolic compounds and their microbial-derived metabolites against uropathogenic *Escherichia coli* in bladder epithelial cell cultures. *Int. J. Mol. Sci.* 16(6), 12119–12130.

de Medeiros, A. K. B., De Melo, L. A., Alves, R. A. H., Barbosa, G. A. S., De Lima, K. C. and Porto Carreiro, A. D. F. 2016. Inhibitory effect of cranberry extract on periodontopathogenic biofilm: an integrative review. *J. Indian Soc. Periodontol.* 20(5), 503–508.

Delfanian, M. and Sahari, M. A. 2020. Improving functionality, bioavailability, nutraceutical and sensory attributes of fortified foods using phenolics-loaded nanocarriers as natural ingredients. *Food Res. Int.* 137, 109555.

Desrouillères, K., Millette, M., Bagheri, L., Maherani, B., Jamshidian, M. and Lacroix, M. 2020. The synergistic effect of cell wall extracted from probiotic biomass containing Lactobacillus acidophilus CL1285, L. casei LBC80R, and L. rhamnosus CLR2 on the anticancer activity of cranberry juice-HPLC fractions. *J. Food Biochem.* 44(5), e13195.

Devore, E. E., Kang, J. H., Breteler, M. M. and Grodstein, F. 2012. Dietary intakes of berries and flavonoids in relation to cognitive decline. *Ann. Neurol.* 72(1), 135–143.

Dini, I. and Laneri, S. 2019. Nutricosmetics: A brief overview. *Phytother. Res.* 33(12), 3054–3063.

Eslam, M., Newsome, P. N., Sarin, S. K., Anstee, Q. M., Targher, G., Romero-Gomez, M., Zelber-Sagi, S., Wai-Sun Wong, V., Dufour, J. F., Schattenberg, J. M., Kawaguchi, T., Arrese, M., Valenti, L., Shiha, G., Tiribelli, C., Yki-Järvinen, H., Fan, J. G., Grønbæk, H., Yilmaz, Y., Cortez-Pinto, H., Oliveira, C. P., Bedossa, P., Adams, L. A., Zheng, M. H., Fouad, Y., Chan, W. K., Mendez-Sanchez, N., Ahn, S. H., Castera, L., Bugianesi, E., Ratziu, V. and George, J. 2020. A new definition for metabolic dysfunction-associated fatty liver disease: an international expert consensus statement. *J. Hepatol.* 73(1), 202–209.

© Burleigh Dodds Science Publishing Limited, 2022. All rights reserved.

Fassarella, M., Blaak, E. E., Penders, J., Nauta, A., Smidt, H. and Zoetendal, E. G. 2021. Gut microbiome stability and resilience: Elucidating the response to perturbations in order to modulate gut health. *Gut* 70(3), 595–605.

FDA. 2020. *Qualified Health Claim for Certain Cranberry Products and Urinary Tract Infections*. FDA. Available from: https://www.fda.gov/food/cfsan-constituent-updates/fda-announces-qualified-health-claim-certain-cranberry-products-and-urinary-tract-infections#:~:text=%E2%80%9CConsuming%20one%20serving%20(8%20oz,claim%20is%20limited%20and%20inconsistent.%E2%80%9D.

Feghali, K., Feldman, M., La, V. D., Santos, J. and Grenier, D. 2012. Cranberry proanthocyanidins: Natural weapons against periodontal diseases. *J. Agric. Food Chem.* 60(23), 5728-5735.

Feliciano, R. P., Mills, C. E., Istas, G., Heiss, C. and Rodriguez-Mateos, A. 2017. Absorption, metabolism and excretion of cranberry (poly)phenols in humans: A dose response study and assessment of inter-individual variability. *Nutrients* 9(3), 268. doi: 10.3390/nu9030268.

Feng, M., Zheng, X., Wan, J., Pan, W., Xie, X., Hu, B., Wang, Y., Wen, H. and Cai, S. 2021. Research progress on the potential delaying skin aging effect and mechanism of tea for oral and external use. *Food Funct.* 12(7), 2814–2828.

Foxman, B. 1990. Recurring urinary tract infection: Incidence and risk factors. *Am. J. Public Health* 80(3), 331–333.

Foxman, B. and Brown, P. 2003. Epidemiology of urinary tract infections: Transmission and risk factors, incidence, and costs. *Infect. Dis. Clin. North Am.* 17(2), 227–241.

Fraga, C. G., Croft, K. D., Kennedy, D. O. and Tomás-Barberán, F. A. 2019. The effects of polyphenols and other bioactives on human health. *Food Funct.* 10(2), 514–528.

Fu, Z., Liska, D., Talan, D. and Chung, M. 2017. Cranberry reduces the risk of urinary tract infection recurrence in otherwise healthy women: A systematic review and meta-analysis. *J. Nutr.* 147(12), 2282–2288.

Furumura, M., Sato, N., Kusaba, N., Takagaki, K. and Nakayama, J. 2012. Oral administration of French maritime pine bark extract (Flavangenol®) improves clinical symptoms in photoaged facial skin. *Clin. Interv. Aging* 7, 275–286.

Gao, C., Cunningham, D. G., Liu, H., Khoo, C. and Gu, L. 2018. Development of a Thiolysis HPLC method for the analysis of procyanidins in cranberry products. *J. Agric. Food Chem.* 66(9), 2159-2167.

Gildawie, K. R., Galli, R. L., Shukitt-Hale, B. and Carey, A. N. 2018. Protective effects of foods containing flavonoids on age-related cognitive decline. *Curr. Nutr. Rep.* 7(2), 39–48.

González de Llano, D., Moreno-Arribas, M. V. and Bartolomé, B. 2020. Cranberry polyphenols and prevention against urinary tract infections: relevant considerations. *Molecules* 25(15), 3523.

Gotteland, M., Andrews, M., Toledo, M., Muñoz, L., Caceres, P., Anziani, A., Wittig, E., Speisky, H. and Salazar, G. 2008. Modulation of *Helicobacter pylori* colonization with cranberry juice and Lactobacillus johnsonii La1 in children. *Nutrition* 24(5), 421–426.

Grammatikopoulou, M. G., Gkiouras, K., Papageorgiou, S. ?, Myrogiannis, I., Mykoniatis, I., Papamitsou, T., Bogdanos, D. P. and Goulis, D. G. 2020. Dietary factors and supplements influencing prostate specific-antigen (PSA) concentrations in men with prostate cancer and increased cancer risk: An evidence analysis review based on randomized controlled trials. *Nutrients* 12(10).

© Burleigh Dodds Science Publishing Limited, 2022. All rights reserved.

Gratton, G., Weaver, S. R., Burley, C. V., Low, K. A., Maclin, E. L., Johns, P. W., Pham, Q. S., Lucas, S. J. E., Fabiani, M. and Rendeiro, C. 2020. Dietary flavanols improve cerebral cortical oxygenation and cognition in healthy adults. *Sci. Rep.* 10(1), 19409.

Greene, A. C., Acharya, A. P., Lee, S. B., Gottardi, R., Zaleski, E. and Little, S. R. 2020. Cranberry extract-based formulations for preventing bacterial biofilms. *Drug Deliv. Transl. Res.* 11(3), 1144–1155.

Gu, L., Kelm, M. A., Hammerstone, J. F., Beecher, G., Holden, J., Haytowitz, D., Gebhardt, S. and Prior, R. L. 2004. Concentrations of proanthocyanidins in common foods and estimations of normal consumption. *J. Nutr.* 134(3), 613–617.

Gu, L., Kelm, M. A., Hammerstone, J. F., Beecher, G., Holden, J., Haytowitz, D. and Prior, R. L. 2003. Screening of foods containing proanthocyanidins and their structural characterization using LC-MS/MS and thiolytic degradation. *J. Agric. Food Chem.* 51(25), 7513-7521.

Guevara, B. and Cogdill, A. G. 2020. Helicobacter pylori: A review of current diagnostic and management strategies. *Dig. Dis. Sci.* 65(7), 1917–1931.

Gupta, K., Hooton, T. M., Naber, K. G., Wullt, B., Colgan, R., Miller, L. G., Moran, G. J., Nicolle, L. E., Raz, R., Schaeffer, A. J., Soper, D. E., Infectious Diseases Society of America and European Society for Microbiology and Infectious Diseases. 2011. International clinical practice guidelines for the treatment of acute uncomplicated cystitis and pyelonephritis in women: A 2010 update by the Infectious Diseases Society of America and the European Society for Microbiology and Infectious Diseases. *Clin. Infect. Dis.* 52(5), e103–e120.

Handog, E. B., Galang, D. A., De Leon-Godinez, M. A. and Chan, G. P. 2009. A randomized, double-blind, placebo-controlled trial of oral procyanidin with vitamins A, C, E for melasma among Filipino women. *Int. J. Dermatol.* 48(8), 896–901.

Hao, W., Kwek, E., He, Z., Zhu, H., Liu, J., Zhao, Y., MA, Ma, K. Y., He, W. S. and Chen, Z. Y. 2020. Ursolic acid alleviates hypercholesterolemia and modulates the gut microbiota in hamsters. *Food Funct.* 11(7), 6091–6103.

Harnly, J. M., Doherty, R. F., Beecher, G. R., Holden, J. M., Haytowitz, D. B., Bhagwat, S. and Gebhardt, S. 2006. Flavonoid content of U.S. fruits, vegetables, and nuts. *J. Agric. Food Chem.* 54(26), 9966–9977.

Hernandez, D. F., Cervantes, E. L., Luna-Vital, D. A. and Mojica, L. 2020. Food-derived bioactive compounds with anti-aging potential for nutricosmetic and cosmeceutical products. *Crit. Rev. Food Sci. Nutr.*, 1–16.

Hidron, A. I., Edwards, J. R., Patel, J., Horan, T. C., Sievert, D. M., Pollock, D. A., Fridkin, S. K., National Healthcare Safety Network Team and Participating National Healthcare Safety Network Facilities 2008. NHSN annual update: Antimicrobial-resistant pathogens associated with healthcare-associated infections: Annual summary of data reported to the National Healthcare Safety Network at the Centers for Disease Control and Prevention, 2006–2007. *Infect. Control Hosp. Epidemiol.* 29(11), 996–1011.

Hormoznejad, R., Mohammad Shahi, M., Rahim, F., Helli, B., Alavinejad, P. and Sharhani, A. 2020. Combined cranberry supplementation and weight loss diet in non-alcoholic fatty liver disease: A double-blind placebo-controlled randomized clinical trial. *Int. J. Food Sci. Nutr.* 71(8), 991–1000.

Ikäheimo, R., Siitonen, A., Heiskanen, T., Kärkkäinen, U., Kuosmanen, P., Lipponen, P. and Mäkelä, P. H. 1996. Recurrence of urinary tract infection in a primary care setting: Analysis of a 1-year follow-up of 179 women. *Clin. Infect. Dis.* 22(1), 91–99.

© Burleigh Dodds Science Publishing Limited, 2022. All rights reserved.

Jepson, R. G., Williams, G. and Craig, J. C. 2012. Cranberries for preventing urinary tract infections. *Cochrane Database Syst. Rev.* 10, CD001321.

Kent, K., Charlton, K. E., Netzel, M. and Fanning, K. 2017. Food-based anthocyanin intake and cognitive outcomes in human intervention trials: A systematic review. *J. Hum. Nutr. Diet.* 30(3), 260–274.

Kesse-Guyot, E., Fezeu, L., Andreeva, V. A., Touvier, M., Scalbert, A., Hercberg, S. and Galan, P. 2012. Total and specific polyphenol intakes in midlife are associated with cognitive function measured 13 years later. *J. Nutr.* 142(1), 76–83.

Koradia, P., Kapadia, S., Trivedi, Y., Chanchu, G. and Harper, A. 2019 Probiotic and cranberry supplementation for preventing recurrent uncomplicated urinary tract infections in premenopausal women: A controlled pilot study. *Expert Rev. Anti Infect. Ther.* 17(9), 733–740.

Kresty, L. A., Howell, A. B. and Baird, M. 2011. Cranberry proanthocyanidins mediate growth arrest of lung cancer cells through modulation of gene expression and rapid induction of apoptosis. *Molecules* 16(3), 2375–2390.

Li, H. Y., Gan, R. Y., Shang, A., Mao, Q. Q., Sun, Q. C., Wu, D. T., Geng, F., He, X. Q. and Li, H. B. 2021. Plant-based foods and their bioactive compounds on fatty liver disease: Effects, mechanisms, and clinical application. *Oxid. Med. Cell. Longev.* 2021, 6621644.

Li, Z. X., Ma, J. L., Guo, Y., Liu, W. D., Li, M., Zhang, L. F., Zhang, Y., Zhou, T., Zhang, J. Y., Gao, H. E., Guo, X. Y., Ye, D. M., Li, W. Q., You, W. C. and Pan, K. F. 2020. Suppression of *Helicobacter pylori* infection by daily cranberry intake: A double-blind, randomized, placebo-controlled trial. *J. Gastroenterol. Hepatol.* 36(4), 927–935.

Liska, D. J., Kern, H. J. and Maki, K. C. 2016. Cranberries and urinary tract infections: How can the same evidence lead to conflicting advice? *Adv. Nutr.* 7(3), 498–506.

Mabeck, C. E. 1972. Treatment of uncomplicated urinary tract infection in non-pregnant women. *Postgrad. Med. J.* 48(556), 69–75.

Maki, K. C., Kaspar, K. L., Khoo, C., Derrig, L. H., Schild, A. L. and Gupta, K. 2016. Consumption of a cranberry juice beverage lowered the number of clinical urinary tract infection episodes in women with a recent history of urinary tract infection. *Am. J. Clin. Nutr.* 103(6), 1434–1442.

Mantzorou, M., Zarros, A., Vasios, G., Theocharis, S., Pavlidou, E. and Giaginis, C. 2019. Cranberry: A promising natural source of potential nutraceuticals with anticancer activity. *Anti Cancer Agents Med. Chem.* 19(14), 1672–1686.

Marini, A., Grether-Beck, S., Jaenicke, T., Weber, M., Burki, C., Formann, P., Brenden, H., Schönlau, F. and Krutmann, J. 2012. Pycnogenol® effects on skin elasticity and hydration coincide with increased gene expressions of collagen type I and hyaluronic acid synthase in women. *Skin Pharmacol. Physiol.* 25(2), 86–92.

Marsh, P. D. 2018. In Sickness and in health-what does the oral microbiome mean to us? An ecological perspective. *Adv. Dent. Res.* 29(1), 60–65.

Mastroiacovo, D., Kwik-Uribe, C., Grassi, D., Necozione, S., Raffaele, A., Pistacchio, L., Righetti, R., Bocale, R., Lechiara, M. C., Marini, C., Ferri, C. and Desideri, G. 2015. Cocoa flavanol consumption improves cognitive function, blood pressure control, and metabolic profile in elderly subjects: The Cocoa, Cognition, and Aging (CoCoA) Study–A randomized controlled trial. *Am. J. Clin. Nutr.* 101(3), 538–548.

Mena, P., González De Llano, D., Brindani, N., Esteban-Fernández, A., Curti, C., Moreno-Arribas, M. V., Del Rio, D. and Bartolomé, B. 2017. 5-(3′,4′-dihydroxyphenyl)-⍰-va lerolactone and its sulphate conjugates, representative circulating metabolites of

© Burleigh Dodds Science Publishing Limited, 2022. All rights reserved.

flavan-3-ols, exhibit anti-adhesive activity against uropathogenic *Escherichia coli* in bladder epithelial cells. *J. Funct. Foods* 29, 275–280.

Michalak, M., Pierzak, M., Kręcisz, B. and Suliga, E. 2021. Bioactive compounds for skin health: A review. *Nutrients* 13(1), 203.

Neto, C. C. 2007. Cranberry and its phytochemicals: A review of in vitro anticancer studies. *J. Nutr.* 137(1), 186S–193S.

Neto, C. C. and Vinson, J. A. 2011. Cranberry. In: Benzie Iff, W.-G. S. (Ed.) *Herbal Medicine: Biomolecular and Clinical Aspects* (2nd edn.). Boca Raton, FL: CRC Press Press/ Taylor & Francis.

Nicolle, L. E. and AMMI Canada Guidelines Committee. 2005. Complicated urinary tract infection in adults. *Can. J. Infect. Dis. Med. Microbiol.* 16(6), 349–360.

Niska, R., Bhuiya, F. and Xu, J. 2010. National hospital ambulatory medical care survey: 2007 emergency department summary. *Natl Health Stat. Rep.* 26, 1–31.

Nobile, V., Michelotti, A., Cestone, E., Caturla, N., Castillo, J., Benavente-García, O., Pérez-Sánchez, A. and Micol, V. 2016. Skin photoprotective and antiageing effects of a combination of rosemary (*Rosmarinus officinalis*) and grapefruit (*Citrus paradisi*) polyphenols. *Food Nutr. Res.* 60, 31871.

Ou, K., Percival, S. S., Zou, T., Khoo, C. and Gu, L. 2012. Transport of cranberry A-type procyanidin dimers, trimers, and tetramers across monolayers of human intestinal epithelial Caco-2 cells. *J. Agric. Food Chem.* 60(6), 1390–1396.

Özcan, E., Sun, J., Rowley, D. C. and Sela, D. A. 2017. A human gut commensal ferments cranberry carbohydrates to produce formate. *Appl. Environ. Microbiol.* 83(17), 3637–3644.

Peres, M. A., Macpherson, L. M. D., Weyant, R. J., Daly, B., Venturelli, R., Mathur, M. R., Listl, S., Celeste, R. K., Guarnizo-Herreño, C. C., Kearns, C., Benzian, H., Allison, P. and Watt, R. G. 2019. Oral diseases: A global public health challenge. *Lancet* 394(10194), 249–260.

Pérez-Sánchez, A., Barrajón-Catalán, E., Caturla, N., Castillo, J., Benavente-García, O., Alcaraz, M. and Micol, V. 2014. Protective effects of citrus and rosemary extracts on UV-induced damage in skin cell model and human volunteers. *J. Photochem. Photobiol. B* 136, 12–18.

Pérez-Sánchez, A., Barrajón-Catalán, E., Herranz-López, M. and Micol, V. 2018. Nutraceuticals for skin care: A comprehensive review of human clinical studies. *Nutrients* 10(4).

Phenol-Explorer. American cranberry. Available at: http://phenol-explorer.eu/contents/food/74 [Accessed April 6, 2021].

Philip, N., Bandara, H. M. H. N., Leishman, S. J. and Walsh, L. J. 2019. Effect of polyphenol-rich cranberry extracts on cariogenic biofilm properties and microbial composition of polymicrobial biofilms. *Arch. Oral Biol.* 102, 1–6.

Philip, N., Leishman, S. J., Bandara, H. M. H. N., Healey, D. L. and Walsh, L. J. 2020. Randomized controlled study to evaluate microbial ecological effects of CPP-ACP and cranberry on dental plaque. *JDR Clin. Trans. Res.* 5(2), 118–126.

Philip, N. and Walsh, L. J. 2019. Cranberry polyphenols: Natural weapons against dental caries. *Dent. J. (Basel)* 7(1), 20.

Polewski, M. A., Krueger, C. G., Reed, J. D. and Leyer, G. 2016. Ability of cranberry proanthocyanidins in combination with a probiotic formulation to inhibit in vitro invasion of gut epithelial cells by extra-intestinal pathogenic *E. coli*. *J. Funct. Foods* 25, 123–134.

© Burleigh Dodds Science Publishing Limited, 2022. All rights reserved.

Pourmasoumi, M., Hadi, A., Najafgholizadeh, A., Joukar, F. and Mansour-Ghanaei, F. 2020. The effects of cranberry on cardiovascular metabolic risk factors: A systematic review and meta-analysis. *Clin. Nutr.* 39(3), 774-788.

Rafsanjany, N., Senker, J., Brandt, S., Dobrindt, U. and Hensel, A. 2015. In vivo consumption of cranberry exerts ex vivo antiadhesive activity against FimH-dominated uropathogenic *Escherichia coli*: A combined in vivo, ex vivo, and in vitro Study of an extract from Vaccinium macrocarpon. *J. Agric. Food Chem.* 63(40), 8804-8818.

Raguzzini, A., Toti, E., Sciarra, T., Fedullo, A. L. and Peluso, I. 2020. Cranberry for bacteriuria in individuals with spinal cord injury: A systematic review and meta-analysis. *Oxid. Med. Cell. Longev.* 2020, 9869851.

Rodriguez-Mateos, A., Feliciano, R. P., Boeres, A., Weber, T., Dos Santos, C. N., Ventura, M. R. and Heiss, C. 2016. Cranberry (poly)phenol metabolites correlate with improvements in vascular function: A double-blind, randomized, controlled, dose-response, crossover study. *Mol. Nutr. Food Res.* 60(10), 2130-2140.

Rodríguez-Morató, J., Matthan, N. R., Liu, J., De La Torre, R. and Chen, C. O. 2018. Cranberries attenuate animal-based diet-induced changes in microbiota composition and functionality: A randomized crossover controlled feeding trial. *J. Nutr. Biochem.* 62, 76-86.

Ronis, M. J. J., Pedersen, K. B. and Watt, J. 2018. Adverse effects of nutraceuticals and dietary supplements. *Annu. Rev. Pharmacol. Toxicol.* 58, 583-601.

Saliou, C., Rimbach, G., Moini, H., McLaughlin, L., Hosseini, S., Lee, J., Watson, R. R. and Packer, L. 2001. Solar ultraviolet-induced erythema in human skin and nuclear factor-kappa-B-dependent gene expression in keratinocytes are modulated by a French maritime pine bark extract. *Free Radic. Biol. Med.* 30(2), 154-160.

Sánchez, M. C., Ribeiro-Vidal, H., Bartolomé, B., Figuero, E., Moreno-Arribas, M. V., Sanz, M. and Herrera, D. 2020. New evidences of antibacterial effects of cranberry against periodontal pathogens. *Foods* 9(2), 246.

Savoldi, A., Carrara, E., Graham, D. Y., Conti, M. and Tacconelli, E. 2018. Prevalence of antibiotic resistance in Helicobacter pylori: A systematic review and meta-analysis in World Health Organization regions. *Gastroenterology* 155(5), 1372-1382.e17.

Schappert, S. M. and Rechtsteiner, E. A. 2011. Ambulatory medical care utilization estimates for 2007. *Vital Health Stat.* 13(169), 1-38.

Scharf, B., Schmidt, T. J., Rabbani, S., Stork, C., Dobrindt, U., Sendker, J., Ernst, B. and Hensel, A. 2020. Antiadhesive natural products against uropathogenic *E. coli*: What can we learn from cranberry extract? *J. Ethnopharmacol.* 257, 112889.

Seneviratne, C. J., Zhang, C. F. and Samaranayake, L. P. 2011. Dental plaque biofilm in oral health and disease. *Chin. J. Dent. Res.* 14(2), 87-94.

Seyyedmajidi, M., Ahmadi, A., Hajiebrahimi, S., Seyedmajidi, S., Rajabikashani, M., Firoozabadi, M. and Vafaeimanesh, J. 2016. Addition of cranberry to proton pump inhibitor-based triple therapy for *Helicobacter pylori* eradication. *J. Res. Pharm. Pract.* 5(4), 248-251.

Shmuely, H., Yahav, J., Samra, Z., Chodick, G., Koren, R., Niv, Y. and Ofek, I. 2007. Effect of cranberry juice on eradication of *Helicobacter pylori* in patients treated with antibiotics and a proton pump inhibitor. *Mol. Nutr. Food Res.* 51(6), 746-751.

Shukitt-Hale, B., Galli, R. L., Meterko, V., Carey, A., Bielinski, D. F., McGhie, T. and Joseph, J. A. 2005. Dietary supplementation with fruit Polyphenolics ameliorates age-related deficits in behavior and neuronal markers of inflammation and oxidative stress. *Age (Dordr)* 27(1), 49-57.

© Burleigh Dodds Science Publishing Limited, 2022. All rights reserved.

Sloan, R. P., Wall, M., Yeung, L. K., Feng, T., Feng, X., Provenzano, F., Schroeter, H., Lauriola, V., Brickman, A. M. and Small, S. A. 2021. Insights into the role of diet and dietary flavanols in cognitive aging: Results of a randomized controlled trial. *Sci. Rep.* 11(1), 3837.

Srinivas, N. R. 2013. Cranberry juice ingestion and clinical drug-drug interaction potentials; review of case studies and perspectives. *J. Pharm. Pharm. Sci.* 16(2), 289–303.

Stamm, W. E. and Hooton, T. M. 1993. Management of urinary tract infections in adults. *N. Engl. J. Med.* 329(18), 1328–1334.

Stapleton, A. E., Dziura, J., Hooton, T. M., Cox, M. E., Yarova-Yarovaya, Y., Chen, S. and Gupta, K. 2012. Recurrent urinary tract infection and urinary *Escherichia coli* in women ingesting cranberry juice daily: A randomized controlled trial. *Mayo Clin. Proc.* 87(2), 143–150.

Stothers, L. 2002. A randomized trial to evaluate effectiveness and cost effectiveness of naturopathic cranberry products as prophylaxis against urinary tract infection in women. *Can. J. Urol.* 9(3), 1558–1562.

Straub, T. J., Chou, W. C., Manson, A. L., Schreiber, H. L. T., Walker, B. J., Desjardins, C. A., Chapman, S. B., Kaspar, K. L., Kahsai, O. J., Traylor, E., Dodson, K. W., Hullar, M. A. J., Hultgren, S. J., Khoo, C. and Earl, A. M. 2021. Limited effects of long-term daily cranberry consumption on the gut microbiome in a placebo-controlled study of women with recurrent urinary tract infections. *BMC Microbiol.* 21(1), 53.

Student, V., Vidlar, A., Bouchal, J., Vrbkova, J., Kolar, Z., Kral, M., Kosina, P. and Vostalova, J. 2016. Cranberry intervention in patients with prostate cancer prior to radical prostatectomy: Clinical, pathological and laboratory findings. *Biomed. Pap. Med. Fac. Univ. Palacky Olomouc Czech Repub.* 160(4), 559–565.

Sun, J., Deering, R. W., Peng, Z., Najia, L., Khoo, C., Cohen, P. S., Seeram, N. P. and Rowley, D. C. 2019. Pectic oligosaccharides from cranberry prevent quiescence and persistence in the uropathogenic *Escherichia coli* CFT073. *Sci. Rep.* 9(1), 19590.

Suvarna, R., Pirmohamed, M. and Henderson, L. 2003. Possible interaction between warfarin and cranberry juice. *BMJ* 327(7429), 1454.

Takahashi, S., Hamasuna, R., Yasuda, M., Arakawa, S., Tanaka, K., Ishikawa, K., Kiyota, H., Hayami, H., Yamamoto, S., Kubo, T. and Matsumoto, T. 2013. A randomized clinical trial to evaluate the preventive effect of cranberry juice (UR65) for patients with recurrent urinary tract infection. *J. Infect. Chemother.* 19(1), 112–117.

Tan, C. S. S. and Lee, S. W. H. 2021. Warfarin and food, herbal or dietary supplement interactions: A systematic review. *Br. J. Clin. Pharmacol.* 87(2), 352–374.

Travica, N., D'cunha, N. M., Naumovski, N., Kent, K., Mellor, D. D., Firth, J., Georgousopoulou, E. N., Dean, O. M., Loughman, A., Jacka, F. and Marx, W. 2020. The effect of blueberry interventions on cognitive performance and mood: A systematic review of randomized controlled trials. *Brain Behav. Immun.* 85, 96–105.

Turner, A., Chen, S. N., Nikolic, D., Van Breemen, R., Farnsworth, N. R. and Pauli, G. F. 2007. Coumaroyl iridoids and a depside from cranberry (Vaccinium macrocarpon). *J. Nat. Prod.* 70(2), 253–258.

Vidlar, A., Vostalova, J., Ulrichova, J., Student, V., Stejskal, D., Reichenbach, R., Vrbkova, J., Ruzicka, F. and Simanek, V. 2010. The effectiveness of dried cranberries (Vaccinium macrocarpon) in men with lower urinary tract symptoms. *Br. J. Nutr.* 104(8), 1181–1189.

Vostalova, J., Vidlar, A., Simanek, V., Galandakova, A., Kosina, P., Vacek, J., Vrbkova, J., Zimmermann, B. F., Ulrichova, J. and Student, V. 2015. Are high proanthocyanidins

© Burleigh Dodds Science Publishing Limited, 2022. All rights reserved.

key to cranberry efficacy in the prevention of recurrent urinary tract infection? *Phytother. Res.* 29(10), 1559–1567.

Walker, E. B., Barney, D. P., Mickelsen, J. N., Walton, R. J. and Mickelsen, R. A., Jr. 1997. Cranberry concentrate: UTI prophylaxis. *J. Fam. Pract.* 45(2), 167–168.

Walsh, J. M., Ren, X., Zampariello, C., Polasky, D. A., Mckay, D. L., Blumberg, J. B. and Chen, C. Y. 2016. Liquid chromatography with tandem mass spectrometry quantification of urinary proanthocyanin A2 dimer and its potential use as a biomarker of cranberry intake. *J. Sep. Sci.* 39(2), 342–349.

Wang, C. H., Fang, C. C., Chen, N. C., Liu, S. S., Yu, P. H., Wu, T. Y., Chen, W. T., Lee, C. C. and Chen, S. C. 2012. Cranberry-containing products for prevention of urinary tract infections in susceptible populations: A systematic review and meta-analysis of randomized controlled trials. *Arch. Intern. Med.* 172(13), 988–996.

Wang, F., Meng, W., Wang, B. and Qiao, L. 2014. Helicobacter pylori-induced gastric inflammation and gastric cancer. *Cancer Lett.* 345(2), 196–202.

Whyte, A. R., Cheng, N., Fromentin, E. and Williams, C. M. 2018. A randomized, double-blinded, placebo-controlled study to compare the safety and efficacy of low dose enhanced wild blueberry powder and wild blueberry extract (ThinkBlue™) in maintenance of episodic and working memory in older adults. *Nutrients* 10(6), 660. doi: 10.3390/nu10060660.

Woźniewicz, M., Nowaczyk, P. M., Kurhańska-Flisykowska, A., Wyganowska-Świątkowska, M., Lasik-Kurdyś, M., Walkowiak, J. and Bajerska, J. 2018. Consumption of cranberry functional beverage reduces gingival index and plaque index in patients with gingivitis. *Nutr. Res.* 58, 36–45.

Wu, X., Xue, L., Tata, A., Song, M., Neto, C. C. and Xiao, H. 2020. Bioactive components of polyphenol-rich and non-polyphenol-rich cranberry fruit extracts and their chemopreventive effects on colitis-associated colon cancer. *J. Agric. Food Chem.* 68(25), 6845–6853.

Xiao, X., Kim, J., Sun, Q., Kim, D., Park, C. S., Lu, T. S. and Park, Y. 2015. Preventive effects of cranberry products on experimental colitis induced by dextran sulphate sodium in mice. *Food Chem.* 167, 438–446.

Yang-Ou, Y. B., Hu, Y., Zhu, Y. and Lu, N. H. 2018. The effect of antioxidants on Helicobacter pylori eradication: A systematic review with meta-analysis. *Helicobacter* 23(6), e12535.

Zare Javid, A., Maghsoumi-Norouzabad, L., Bazyar, H., Aghamohammadi, V. and Alavinejad, P. 2020. Effects of concurrent Omega-3 and cranberry juice consumption Along with standard antibiotic therapy on the eradication of *Helicobacter pylori*, gastrointestinal symptoms, some serum inflammatory and oxidative stress markers in adults with *Helicobacter pylori* infection: A study protocol for a randomized controlled trial. *Infect. Drug Resist.* 13, 3179–3185.

Zhang, L., MA, Ma, J., Pan, K., Go, V. L., Chen, J. and You, W. C. 2005. Efficacy of cranberry juice on Helicobacter pylori infection: A double-blind, randomized placebo-controlled trial. *Helicobacter* 10(2), 139–145.

Zhao, S., Liu, H. and Gu, L. 2020. American cranberries and health benefits – An evolving story of 25 years. *J. Sci. Food Agric.* 100(14), 5111–5116.

Zhao, S., Zhang, L., Yang, C., Li, Z. and Rong, S. 2019. Procyanidins and Alzheimer's disease. *Mol. Neurobiol.* 56(8), 5556-5567.

© Burleigh Dodds Science Publishing Limited, 2022. All rights reserved.

Chapter 5

Optimizing plant growth, yield and fruit quality with plant bioregulators

Duane Greene, University of Massachusetts, USA

1 Introduction

Plant bioregulators (PBRs) influence many processes in a plant including shoot growth, branch angle, fruit abscission, fruit ripening, fruit shape and fruit finish, bud break and flower bud formation. They function in this capacity as signal givers that lead to gene expression, gene activation or repression. One could view PBRs as the first domino in a series of events that ultimately results in a physiological response. They don't play a direct role or act as a component in a chemical reaction. Rather, they initiate responses to signal gene expression, activation or downregulation and these actually bring about the responses that we attribute to PBR control of processes within the plant.

PBRs are a group of organic compounds that, in small amounts, promote, inhibit or otherwise affect biological and biochemical processes in plants. There are three subcategories of compounds that are generally considered PBRs. First, there are endogenous plant hormones that are naturally occurring, synthesized by the plant. They are intimately involved in the regulations and control of all physiological processes important for regulating growth and development in a plant. There are also endogenous compounds, produced by some, but not all plants, that can regulate specific physiological functions.

http://dx.doi.org/10.19103/AS.2018.0040.09
© Burleigh Dodds Science Publishing Limited, 2019. All rights reserved.

These compounds are not found in all plants and have limited influence on growth and development. PBRs in this category may influence a plant even if they are not produced by that plant. Aminoethoxyvinylglycine (AVG) is an important PBR that falls into this category. There is a third group of organic compounds that are not produced by the plants, but they are active in regulating some function(s) within a plant. They are frequently referred to as synthetic PBRs. All plant hormones are considered PBRs but not all PBRs can be considered as plant hormones.

All aspects of fruit production have undergone many changes in the past 50 years. Among the components that have undergone the greatest change are development and use of new rootstocks, change in cultivars grown, pest management strategies, tree training and pruning systems. A prerequisite for all successful fruit growers is to have the ability to understand and embrace the appropriate innovations that give growers the best chance for success. Even if a grower does adopt all of these innovative practices, growers will not be able to achieve the greatest yield, highest fruit quality, the best pack-out and the highest return for harvested fruit if PBRs are not part of and incorporated into their production and management system. PBRs are unlikely to measurably improve the production of high-quality fruit unless sound management practices are applied to a well-designed orchard.

PBRs are used more extensively in tree fruit production than in any other horticultural or agricultural commodity, and they are essential for effective and profitable production of high-value horticultural crops. They are expensive so they must add significant value to the crop just so that a grower can retrieve the cost of material. The increased value is usually not measured in higher yield but rather in increased value of the harvested crop. This enhancement of value is often manifest in larger fruit size, improved red colour, improved fruit appearance or enhanced fruit quality at harvest or following storage.

The first group of plant hormones to be discovered was the auxins. The structure of indole-3-acetic acid (IAA) was confirmed in the 1930s and similar synthetic compounds that had biological activity that could significantly improve and aid the production of tree fruit were identified. Over the next several decades new PBRs were identified and they became commercially important. There was a lack of consistency, especially with the chemical thinning PBRs that were characterized by an inconsistent response even when used under very similar conditions. In the 1980s more effort was focussed on determining the mode of action of many of these growth regulating compounds. As a result, it was possible to more precisely and consistently use these compounds because of the greater understanding of how these PBRs actually participated in specific physiological events. A better understanding of how PBRs work allows more consistent and targeted use of these compounds in the orchard.

© Burleigh Dodds Science Publishing Limited, 2019. All rights reserved.

PBRs generally require approval and registration for use, and not all are universally available in every country. Although they may require governmental approval for use they should not be considered as pesticides. In some cases PBRs do serve a dual function as PBR and pesticide or at least function to reduce the amount of pesticide that may be required. For example, prohexadione-calcium (Pro-Ca) is a compound that was registered to inhibit vegetative growth on some tree fruit. It limits vegetative growth by inhibiting gibberellin biosynthesis. In the process of growth retardation it alters the structure of the growing shoot which in turn makes the stem less susceptible to invasion of the highly devastating fire blight disease (*Erwinia amylovora*). It functions as one of the most effective methods to control shoot blight rather than having to depend on antibiotic spray, where resistance can be developed if excessively used. Growth retardants can reduce shoot growth in a tree leading to less interior shading and faster drying, thus reducing disease pressure. Reduced shoot growth also allows for better penetration of pesticide sprays into the centre of the tree and reduced use of pesticides.

In this chapter the most important commercially used PBRs will be discussed. Further, how they are used to influence many important orchard activities and how they contribute to and play an important role in the sustainable production of apples will be elaborated.

2 Classification of PBRs

The growth of plants is directed and synchronized by plant hormones (Davies, 1995). These molecules are signal givers that essentially regulate every function of a plant. Not all plant cells respond to hormones and often they respond at only specific stages in the life of a cell. In the absence of plant hormones, a cell would probably remain undifferentiated. Plants can regulate their own synthesis, longevity and destruction as well as affect levels of their hormones in a plant. There are five major classes of hormones that are universally recognized as plant hormones.

2.1 Endogenous plant hormones

2.1.1 Auxins

The auxins were the first group of hormones to be discovered. Since early investigators were unaware of the existence of compounds within a plant that could direct growth and development, the discovery of the first hormone, indole-3-acetic acid (IAA), was a long process that spanned nearly 50 years (Jacobs, 1979). The structure of IAA was identified in the 1930s by Japanese investigators and was recognized as the dominant auxin present within plants. Other auxins were soon identified that were not naturally occurring including

© Burleigh Dodds Science Publishing Limited, 2019. All rights reserved.

naphthaleneacetic acid (NAA) and indole-3-butyric acid (IBA). Research soon revealed that auxins played an important role in many plant processes including apical dominance, phototropism, fruit growth, fruit set, root initiation, fruit ripening and leaf abscission. Synthetic auxins, not the naturally occurring IAA, were more active in influencing physiological process and thus they were adopted and developed for commercial use.

2.1.2 Gibberellins

Unlike auxins which were dominated primarily by one compound, the gibberellins include a very large group that numbers in the dozens. Although there are many endogenous gibberellins identified, within a specific type of plant (e.g. apple) there are just a few that are dominant in the gibberellin pool (Cleland, 1969). For example, in apple, gibberellins A_4 and A_7 are present in the highest concentration (Phinney, 1983). The physiological activity of gibberellins differs widely. A gibberellin may be highly active in one type of plant yet inactive in another plant. Gibberellins are recognized for being very important in regulating terminal growth in plants. However, stimulation of terminal growth is not one of the commercial uses of gibberellins. More frequently, plants are treated with plant bioregulators that specifically inhibit gibberellin biosynthesis. Paclobutrazol, uniconazole and prohexadione-calcium (Pro-Ca) are examples of gibberellin biosynthesis inhibitors used to reduce plant growth. Each differ in the point along the gibberellin biosynthetic pathway where they inhibit the synthesis. Gibberellins can inhibit flowering in trees. In pome fruit, gibberellins produced by the seeds diffuse out of the seeds, migrate to growing points where flower bud formation occurs and inhibit the meristem from developing into a flower cluster. Gibberellins are an important hormone group that facilitate and help direct seed germination.

2.1.3 Cytokinins

In the 1940s studies were initiated to identify compound(s) that were responsible for increased growth of leaf disks. It was determined that an adenine-based compound present in yeast extract was involved. Carlos Miller in Folk Skoog's lab in Wisconsin fractionated yeast extract and herring sperm DNA and found cell division activity using a tobacco pith assay that was perfected there (Miller et al., 1955). The cell division factor was identified as an amino purine and it was called kinetin, because it stimulated strong cell division (cytokinesis) activity. Kinetin was not a naturally occurring compound but rather a breakdown product of the herring sperm DNA. Many analogs of kinetin were synthesized and activity confirmed using the tobacco callus assay. Likely structures were identified and efforts were then directed to identify a similar compound that

© Burleigh Dodds Science Publishing Limited, 2019. All rights reserved.

may be present in plants. Letham (1963) isolated a very active compound from immature corn kernels and called this compound zeatin. The primary cytokinins found in plants are zeatin and zeatin riboside, although there are a few other cytokinins, with similar structures present in lower concentrations, that do possess cytokinin activity.

Cytokinins are associated with tissue undergoing rapid cell division. They interact with auxins to determine the morphological development of undifferentiated cells. A high ratio where auxin is favoured results in tissue developing into roots whereas changing the ratio by increasing the cytokinins will favour the growth of the cells into shoots. There is strong evidence that a major site of cytokinin synthesis in plants is the roots and the primary direction of movement in the plant is upward through the xylem.

2.1.4 Abscisic acid

Experiments that led to the discovery of abscisic acid (ABA) were conducted independently in two different laboratories in the 1960s (Moore, 1989). Addicott and his group were studying hormonal relationship in cotton. They identified a compound that had strong abscission promoting properties and they called it abscisin II. Wareing's group in Wales was studying dormancy in trees and identified a compound that promoted dormancy in woody plants. Later it was determined that these two compounds were identical and had the same structure leading to the adoption of the name abscisic acid (Milborrow, 1984). While it is true that ABA can cause leaf abscission, it is more prominently associated with the onset of dormancy in plants, and for playing a dominant role in the regulation of water relations in a plant by regulating the stomatal functions. When a plant is stressed, ABA level increases and this signals the pumping of cations out of the guard cells which results in osmatic movement of water from the guard cell, closing the stomates. When the leaves rehydrate ABA decreases in the leaf, cations are pumped back into the guard cells, causing them to become turgid and thus opening the stomates.

2.1.5 Ethylene

The hormonal properties of ethylene were first recognized by the Russian scientist Nejebulov in 1901 (Beyer et al., 1984). He showed that illuminating gas could cause leaf abscission and epinasty. Several scientists over the next three decades made discoveries that showed that ethylene had hormone-like activity. However, it was Crocker, a post-harvest physiologist, who first suggested that ethylene was an endogenous plant hormone (Crocker et al., 1935). Few scientists agreed with Crocker since most had a difficult time assigning ethylene a status as a legitimate endogenous hormone with important regulatory activity, since

© Burleigh Dodds Science Publishing Limited, 2019. All rights reserved.

it was a gas and could move freely in the air and in the plant (Reid and Howell, 1995). It was not until the 1960s when the improved analytical properties of the gas chromatograph were perfected that scientists were able to easily measure small quantities of ethylene which then allowed them to recognize and acknowledge the hormonal properties of ethylene. Ethylene is unique among the plant hormones in that it can move within the intracellular space of a plant as a gas, dissolve in the cytoplasm and act as the precursor of ethylene, 1-amino-cyclo-propane-1-carboxylic acid (ACC). Ethylene is a dominant hormone involved in the ripening process. Ripening of fruit and the post-harvest life of many fruit and vegetables can be extended by regulating the production or the action of ethylene. Ethylene plays a central role in the abscission process and in some crops can reduce the fruit removal force to facilitate mechanical harvest of some fruit. It can act as a growth retardant and enhance flower bud formation in pome fruit, even in the absence of reducing crop load.

2.2 Minor (lesser known) plant hormones

There are a few other naturally occurring compounds that are considered by many to be plant hormones. While these compounds may not be involved in many physiological processes, they are compatible with the definition of a plant hormone and therefore they are included in the plant hormone category.

2.2.1 Jasmonates

Jasmonic acid (JA) and methyl jasmonate (MeJA) are a class of oxylipins composed of a cyclopentane ring with two substituent groups. JAs and their derivatives activate a succession of signalling pathways that can influence aspects of fruit ripening, production of viable pollen, root growth and plant resistance to insect and pathogen invasion (Creelman and Mullet, 1997). Following mechanical wounding JA biosynthesis is increased which in turn signals the expression of the appropriate gene response. Jasmonates appear to be involved in the early steps of the ripening of climacteric fruit in conjunction with ethylene (Fan et al., 1998). JA may also enhance the production of anthocyanins (Rudell et al., 2005). There is a JA proprietary formulation that is currently being used to enhance red colour development of apples when applied within 4 weeks prior to harvest.

2.2.2 Brassinosteroids

The brassinosteroids were first discovered in the 1970s in pollen grains of rape (*Brassica napus*). Since that time they have been found in most other structures in a plant, but in very low concentrations. Over 70 brassinosteroids have been

© Burleigh Dodds Science Publishing Limited, 2019. All rights reserved.

discovered. They have the basic steroid nucleus of four fused rings and an alkyl side chain, but differing isomeric configuration and substituent groups. They are involved in cell expansion (Clouse, 1996), vascular differentiation (Fukuda, 1997), cell elongation (Katsumi, 1991) and others (Savaldi-Goldstein, 2006). Brassinosteroids are required for growth and reproduction (Clouse and Sasse, 1998). Brassinosteroids have shown promise for agricultural use, but the extremely low levels found within a plant have slowed progress.

2.3 Plant growth regulators

Perhaps the most important PBRs in general use in tree fruit production are compounds that are not produced by the plant or if they are, their occurrence is not widespread. Most are chemically synthesized similar to many biochemicals while others are produced by fermentation.

2.3.1 Aminoethoxyvinylglycine (ReTain®)

AVG is a compound that was identified as an ethylene biosynthesis inhibitor (Boller et al., 1979). Hoffman-La Roche was the first company to attempt to commercialize this compound. There was a flurry of research activity that confirmed that it had strong biological activity especially when used for delaying senescence. However, development of AVG was abandoned because of the high cost of production. While other companies looked at the compound, all lost interest and AVG remained undeveloped. In the late 1980s, registration of an important pre-harvest drop control compound for apples (daminozide) was dropped, leaving a void in the market. Bangerth (1978) previously showed that AVG could control pre-harvest drop of apples. Abbott Laboratories started to evaluate this compound for commercial use as a drop control compound on apples as well as for other uses. Following several years of development and testing AVG received full label registration in 1997 as ReTain® to retard pre-harvest drop, aid in harvest management and to improve fruit quality. It took nearly 20 years from the time the pre-harvest drop control activity of AVG was recognized in late 1970s until it was registered as a commercial product.

2.3.2 6-Benzyladenine

The road from discovery, through evaluation, field testing and marketing of 6-benzyladenine (6-BA) occurred over a 35-year period. There were three reasons for this extremely long development period (Greene et al., 2016). First, 6-BA is a cytokinin and abscission was not one of the physiological characteristics attributed to cytokinins. Second, the apple industry is constantly undergoing

© Burleigh Dodds Science Publishing Limited, 2019. All rights reserved.

change. As the industry evolved (training systems, rootstock, cultivars etc.) new needs emerged, which in turn created the opportunities for new uses for existing PBRs. Third, there were real challenges posed in formulating and packaging BA as a registrable product that required time and formulation research to resolve. The ability of 6-BA and the gibberellins A_{4+7} to increase the length of 'Delicious' apples and to cause bud to break on apple stems was recognized by Williams and Stahly (1969). Abbott laboratories combined these two compounds into a proprietary product (Promalin®) that could both elongate 'Delicious' and cause bud break on young apple shoots (Williams and Billingsley, 1970). When this product was applied to young trees in the orchard to improve branching structure, it was found that, at the rates used it cause fruit thinning. It was determined that it was the 6-BA that was the main component that was causing the thinning (Greene et al., 2016). Rather than develop a new 6-BA-alone product for chemical thinning, Abbott Laboratories altered their Promalin® product by significantly reducing the GA component. All indications from the literature suggested that the small amount of GA in the formulation should have no effect on the abscission capability of the 6-BA. However, this newly formulated product did not thin trees as well and on treated trees there often remained clusters of small fruit that had reduced seed numbers. A new 6-BA product was then formulated, tested and registered for use as the product MaxCel® in 2004.

2.3.3 2-Chloroethylphosphonic acid (ethephon, Ethrel®)

In the 1960s ethylene was recognized as an important hormone that could influence many physiological processes. Since it is a gas in its natural state, it is difficult to administer it to plants unless they are confined in an enclosed space. Activities such as ripening bananas, avocado and tomatoes and degreening of citrus are activities compatible with treatment in enclosed rooms (Barry and Giovannoni, 2007). However, ethylene is very active in causing ripening, fruit abscission and other physiological responses where only application in the orchard or field is practical. In the 1960s a compound was identified by scientists at Amchem Products Inc. that would produce ethylene gas when exposed to alkaline conditions. This product was identified as Amchem 66-329 and eventually was given the common name of ethephon (Ethrel®) (Yang, 1969). It was stable under acid conditions and when sprayed on a plant it could be taken up by a plant. When it was absorbed into the cell where the cytoplasm was alkaline or had a neutral pH, it broke down and in the process ethylene gas was liberated in the plant. Over the past five decades ethephon has been the compound of choice for treating trees in the orchards and fields with ethylene. There are many commercial uses of this product in fruit production including vegetative growth control, enhanced flower bud formation, advanced ripening

© Burleigh Dodds Science Publishing Limited, 2019. All rights reserved.

of fruit, reduce fruit removal force to aid mechanical harvesting and chemically thinning of apples.

2.3.4 Naphthaleneacetic acid

The auxins were the first group of hormones to be discovered and identified and NAA is one of the few auxins that still retains registration for use. It is a very important fruitlet thinner used on apples and it is also an effective pre-harvest drop control compound. When applied to plants it can stimulate ethylene production. Therefore, it is not uncommon for plants treated with NAA to display symptoms of ethylene production such as epinasty of leaves and shoots and advancement of fruit ripening.

2.3.5 Prohexadione-calcium (Pro-Ca, Apogee®, Kudos®)

The spacing between trees in contemporary orchards is being reduced resulting in orchards with less space between trees. Regulation and control of the vegetative growth in these orchards is becoming more important. The use of growth retardants is a strategy used by many growers to reduce vegetative growth and minimize shading within the orchard. It is a gibberellin biosynthesis inhibitor that was introduced in 1997. It can effectively control growth while having few negative side effects. This compound is the product of choice and the primary growth retardant used to manage vegetative growth in apples.

3 Application of PBRs

3.1 Foliar sprays

The majority of PBRs are applied as foliar sprays. The effectiveness of the application is generally gauged by observing or quantifying a physiological response as a result of the application. However, the spray application details and the environmental factors associated with the application may influence the uptake of the PBR and thus play an important role in determining the ultimate response to the PBR.

Traditionally, foliar sprays of PBRs were made as a dilute application where sprays were applied to the drip point. This approach is less common now due in part to loss of spray by runoff, compaction of the soil as a result of many trips through the orchard and increased fuel consumption which are not sustainable practices. Further, valuable time is lost by spending unproductive time driving through the orchard and having to fill the sprayer tank frequently. Further, the makeup and design of orchards has changed dramatically in recent years with the widespread adoption and use of dwarfing rootstocks and more efficient

© Burleigh Dodds Science Publishing Limited, 2019. All rights reserved.

training systems which provided more attractive options for efficient application of plant growth regulators (PGRs).

The trend of the past few decades is to plant trees at much higher densities. The increase in orchard density has led to a reduction in the size of application equipment needed and a reduction in the volume of spray required to deliver pesticides and PBRs. Byers et al. (1971) recognized the problem and they were the first to introduce the concept of tree-row volume (TRV) to determine the volume of spray necessary to achieve adequate coverage of trees planted at different tree densities. An example of a dilute TRV calculation for an apple orchard is illustrated in Table 1. The concept has evolved and has been redefined over time. Effective and consistent use of PBRs must take TRV into account if spray applications of PBRs are to be applied at reduced spray volumes (Bukovac, 1980). For example, label recommendations for the commonly used PBRs, such as 6-BA, NAA or AVG, initially made label recommendations based upon grams active ingredient per ha. Wider acceptance of the TRV concept has led to recent label modifications which result in recommendations that use mLs or grams of product per volume of spray per ha.

The response to PBRs is generally rate sensitive. If too little is applied there is either no response or an insufficient response. If excess is applied there could be detrimental responses such as phytotoxicity or over responses. In developing protocols for grower use, it is important to assure that the correct amount of material is applied to achieve the desired response. This is relatively

Table 1 Calculations of dilute volume necessary to wet apple foliage in an orchard to saturation

Parameters to measure
Canopy height (m) - distance from first scaffold to the top of the canopy
Tree width (m) - average maximum width of a tree from branch tip to branch tip

[illegible]

[illegible]

Dilute volume requirement per ha equals about 1 L of spray for each 10 m^3 of foliage

Example calculation of dilute spray requirement
Trees in a sample orchard have a canopy height of 4 m, a tree width of 3 m and a row width of 5 m

[illegible]

[illegible]

© Burleigh Dodds Science Publishing Limited, 2019. All rights reserved.

easily accomplished by making a dilute application since all parts of the tree receive the desired dose and any excess will drip off.

However, application of pesticides and PBRs is generally done now using concentrate spray application. Concentrate spraying involves applying the desired and calculated amount of active ingredient on an orchard using a reduced amount of water. In order to do this correctly, it is first necessary to calculate the TRV of the orchard to be sprayed. From this you can calculate the amount of material suggested on the label for the area to be sprayed. For example if a block has a TRV of 3740 L ha^{-1} (400 gal $acre^{-1}$) and the grower decides to apply a PBR at a 2X concentration or using half the amount of water. The grower would then put in the sprayer tank the amount of material suggested to make the dilute application on that block. The block will then be sprayed with 1870 L of the water but it would also receive the same amount of active ingredient as the same block sprayed with a larger volume of water.

There are several problems associated with the application of PBRs in a low-volume spray that may result in variable responses (Bukovac, 1985). The possibility of over- or under-application is increased because the response to PBRs is generally linear. For example, if an error of 10% is made in calculation or measurement when making a dilute application there is likely to be a small or minimal influence on the final result since the error is only 10%. However, if the same 10% error is made in 6X concentration application this error will then be increased to 600% which will most likely have a significant and important influence on the final result. When a lower volume of water is used there is generally a faster drying time which is associated with reduced uptake. Finally, small spray particles are associated with concentrate spraying and these are the particles that are most likely to drift and the ones that are most difficult to reach the tops of trees.

Translocation of PBRs is generally limited. Therefore, it is important that the spray application be made directly to the target tree part and does not depend on translocation for more uniform distribution. For example, when AVG is applied to a tree to control pre-harvest drop maximum control is achieved when both the leaves and fruit receive the spray (Fig. 1). If only the fruit received the spray, little drop control was achieved. For maximum response to a PBR precise and directed application is necessary.

3.2 Environmental factors influencing foliar penetration

The plant response to foliar applied growth substances is frequently variable, even when that application is made under, what appears to be, similar or identical application conditions. Environmental factors can cause this variability in several ways.

© Burleigh Dodds Science Publishing Limited, 2019. All rights reserved.

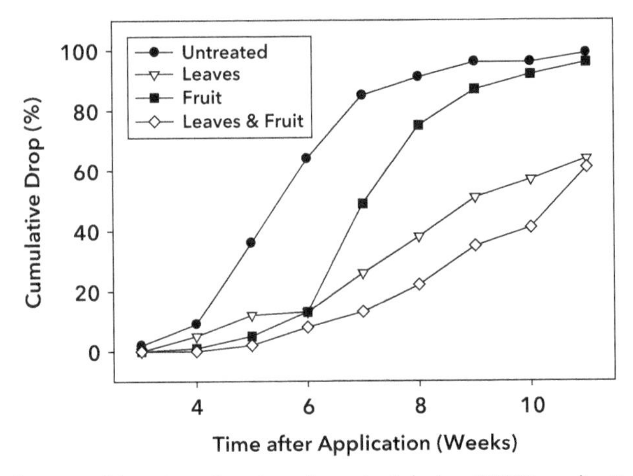

Figure 1 Influence of location of aminoethoxyvinylglycine (AVG) application on apple spurs on the extent of pre-harvest drop on 'McIntosh' apples. Application of AVG made directly to the fruit had minimal influence on pre-harvest drop whereas if the application was made to leaves or leaves plus fruit, pre-harvest drop was controlled for an extended period of time. Source: Greene (2006).

3.2.1 Environmental conditions prior to application

Uptake into the leaf must occur through the cuticle which is present on both the upper and lower leaf surfaces. During leaf development, epicuticular waxes are exuded from underlying cells onto the leaf surfaces and these influence uptakes into the leaf. It is generally accepted that the amount, structure and composition of epicuticular waxes influence leaf wettability and penetration of foliar applied substances (Bukovac, 1973). During cool, moist, cloudy periods leaf cuticle development is slowed whereas during hot, dry sunny weather the cuticle development is more robust. It has been observed in the field that response to a foliar applied substance is greater when leaves develop under cool, cloudy conditions and a reduced response is seen when leaves develop under sunny, dry and low humidity conditions. The difference in response may be attributed to the amount of PBR taken up by the trees whose leaves were exposed to these different weather conditions while developing.

3.2.2 Environmental conditions during and immediately following application

There are two major factors that influence the uptake of a growth substance into a leaf: temperature and droplet drying time (Greene and Bukovac, 1971). As temperature increases there is a linear increase in uptake of foliar applied substances as illustrated by penetration of NAA through the upper surface of 'Bartlett' pear leaves (Fig. 2). At 25–27°C there is an inflection point where the

© Burleigh Dodds Science Publishing Limited, 2019. All rights reserved.

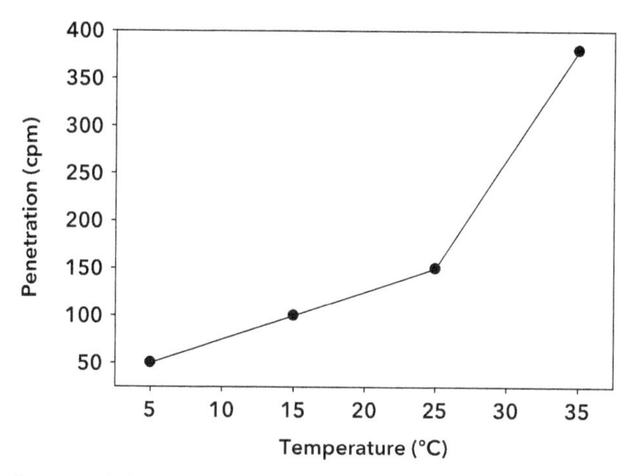

Figure 2 Influence of the temperature at which naphthaleneacetic acid (NAA) was applied on the penetration through the upper surface of 'Bartlett' pear leaves. Penetration of NAA was lower as the temperature increased to 25°C. There was an increase in the slope of update above 25°C that is attributed to a phase change in the cuticle. Source: Greene and Bukovac (1971).

rate of uptake substantially increases. This is attributed to a phase change that is occurring in the cuticle (Price, 1982). This rate in change of uptake makes it somewhat difficult to predict accurately the amount of substance that is being taken up by a leaf. Growers are cautioned to not apply PBRs or to apply these with caution when the temperature approaches 30°C because of the uncertainty associated with the amount of growth substances that are taken up at and above this temperature and consequently to the response of the applied PBR.

The time required for a spray droplet to dry can influence the amount of growth substance that is taken up. This can be illustrated by the penetration of NAA into the 'Bartlett' pear leaf discs when applied to the upper surface (Fig. 3). In general, when a spray is first applied there is a linear increase in penetration of a substance into a leaf that is attributed to diffusion. As a droplet starts to dry the concentration of the growth substance within a drop increases leading to an acceleration in penetration of the growth substance into a leaf. Once the droplet dries there is little or reduced additional penetration. As long as the spray droplet stays intact and does not dry, penetration may occur somewhat continuously (Greene and Bukovac, 1971). The enhancement in uptake will always occur as the droplet dries due largely to the increased concentration of the PBR in the droplet. Therefore, if droplet drying can be slowed, there will be greater uptake of the growth substance into the leaf. Frequently, PBR labels suggest application under high humidity conditions with the express purpose of increasing foliar penetration. In experiments in the more humid Eastern US, when application was made during the day when the temperature was warm

© Burleigh Dodds Science Publishing Limited, 2019. All rights reserved.

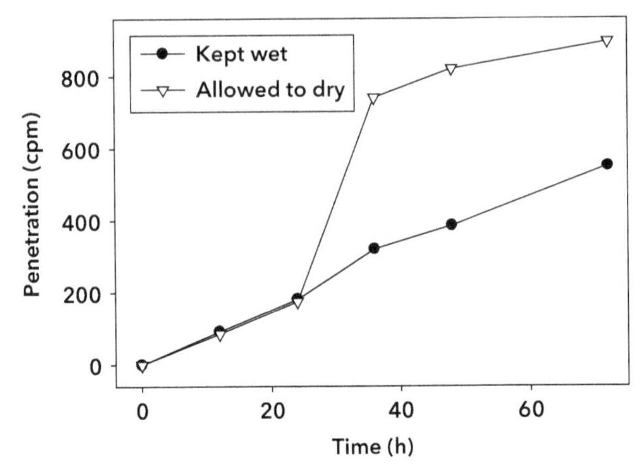

Figure 3 Influence of droplet drying on penetration of naphthaleneacetic acid (NAA) on penetration through the upper surface of 'Bartlett' pear leaves. Penetration is generally linear as long as the spray droplet remains hydrated. However, when the spray droplet starts to dry there is a rapid increase in penetration which tends to stop after the drop dries. Source: Greene and Bukovac (1971).

and drying time rapid it was similar to the uptake when the same amount of growth substance was applied at night when the temperature was cool and drying time was longer (Stover and Greene, 2005). It was concluded that the enhancement in penetration under warm conditions where drying time was shorter was offset by uptake over a longer period of time when the temperature was cooler.

3.3 Influence of rain on PBR response

Frequently, there is a narrow window of opportunity to apply PBRs to be effective. The weather is an uncontrollable factor and predictions are tenuous, especially when it comes to forecasting the probability of rain. Orchardists are often forced, out of necessity, to take chances and apply sprays even if there is a possibility of rain. Sometimes rain does occur following application. The concern then is, just how much activity was lost due to wash-off from the rain? The answer to this question is determined by the time after application (droplet drying) the rain occurred, how much rain occurred, what the solubility properties of the PBR are, the PBR formulation and whether a surfactant was included in the spray. In studies with NAA application on apples, Westwood and Batjer (1960) concluded that some activity was lost if rain occurred within 2 h of application but not if rain occurred after 6–8 h. It appears that this observation may depend on the compound applied and the formulation of that compound. Greene et al. (2000) reported that some loss of pre-harvest drop control activity

© Burleigh Dodds Science Publishing Limited, 2019. All rights reserved.

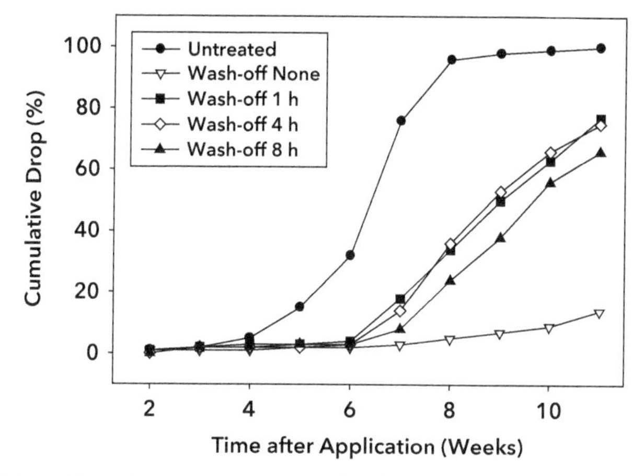

Figure 4 'McIntosh' apple trees were sprayed with aminoethoxyvinylglycine (AVG) that contained no surfactants. Illustrated is the influence of the time after application that trees were thoroughly washed by a high volume of water using a handgun on pre-harvest drop. When trees were sprayed 1-8 h after AVG application, a significant loss in pre-harvest drop control was noted. Source: Greene et al. (2000).

occurred on 'McIntosh' apple trees if AVG was washed off within 24 h after application (Fig. 4) but if the organosilicone surfactant L-77 was included, no activity was lost even if wash-off was done within 1 h after the spray had dried (Fig. 5). Results from field observations made over the years suggest that if the

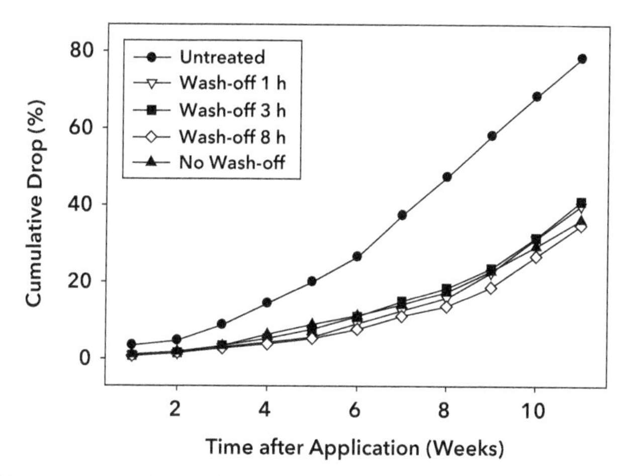

Figure 5 Mature 'McIntosh' apple trees were sprayed with aminoethoxyvinylglycine (AVG) containing 0.01% Silwet L-77 silicone surfactant. Illustrated is the influence of the time after application that trees were thoroughly washed by a high volume of water from a handgun sprayer. Regardless of the time after application pre-harvest drop control was not reduced by thoroughly washing the trees. Source: Greene et al. (2000).

© Burleigh Dodds Science Publishing Limited, 2019. All rights reserved.

spray droplets have completely dried before rain starts, a substantial amount of PBR activity is retained. A rule-of-thumb, which has not been substantiated by research but arrived at through experience and observation, is that if the spray has been dry for at least 30 min prior to the rain event, 80% of the PBR activity will be retained. Sometimes a small amount of rain occurs following application. In this case there is a possibility of enhancing activity rather than inhibition of activity. Bukovac (1965) applied the auxin chlorophenoxyacetic acid to peach leaves. If the leaves were rewetted twice there was a 17% increase in penetration and if the leaves were rewetted five times there was a 47% increase in penetration into the leaf. Field observation following rewetting of the blossom thinners such as dinitrocresol and ammonium thiosulphate (ATS) revealed increased phytotoxicity may be associated with the rewetting of these compounds, which is attributed to increased penetration.

4 Development and maintenance of tree structure

The makeup of apple orchards has changed dramatically over the past half century, especially the tree structure, the rootstocks that the trees are propagated on and the planting systems used. Regardless of the system or the component parts, a common concern expressed by all orchardist is that trees should be encouraged to develop a structure that will allow early and high production. In the 1970s and 1980s the predominant tree structure was a central leader tree propagated on a semi-dwarfing rootstock. The goal in training young trees was to encourage the development of lateral branches that were robust enough to serve as permanent or semi-permanent scaffold branches or develop into sturdy lateral shoots on scaffold branches. Heading cuts (cutting into 1-year-old wood) and the use of a proprietary mixture of 6-BA and gibberellins $A_4 + A_7$ (GA_{4+7}) at rates of 400–500 mg L^{-1} were common. In recent years the tree architecture in the apple industry has evolved rapidly and trees are now propagated primarily on dwarf rootstocks that are spaced very close together. The type of lateral branches that are desirable and are most productive are shorter and weaker shoots.

Most orchardists now order and hope to receive well-feathered trees from the nursery. These trees are the most productive, and come into full production earlier because these branches can produce a small crop as early as the next year. Frequently, trees arrive from the nursery having few if any lateral shoots, or if a few laterals are present, they are often not useful and are cut off. An option available to growers is to apply branch promoting PBR treatments to improve branch development.

Generally, attempts to increase branching on 1-year-old trees the first year in the ground have not been very successful (Forshey, 1982; Miller, 1988). The primary reason for the lack of success is that these trees do not have a

© Burleigh Dodds Science Publishing Limited, 2019. All rights reserved.

well-established and active root system to support the lateral shoot growth desired. Branching successes reported on these types of trees has generally been on very healthy trees that have been planted early and trickle irrigation was installed and used early in the season. Lack of success at increasing branching on first-year trees is common on trees whose growth is compromised by late planting, lack of water, poor soil preparation, poor weed control or inadequate cultural management. After planting it is not uncommon to have sections of the leader that have blind wood. This is especially true with some cultivars that are particularly prone to blind wood: Granny Smith, Fuji, Jonagold and others (Miller, 1988). In general, it is easier and more predictable to stimulate lateral buds to grow when practised on trees the year after planting or on buds located on 2-year-old wood. These trees have an established and growing root system and the attempts to increase lateral branching can be initiated very early in the season when greater success is generally achieved.

4.1 Branching using physical techniques

There are physical techniques to increase branching on young trees including notching, heading, bending the central leader and pinching out very young leaves in the apex. All have disadvantages that have led to lack of extensive use. Of the physical techniques available, notching appears to be the most frequently used technique to improve branching on young apple trees (Greene and Autio, 1994). Notching involves removing a thin piece of bark immediately above an inhibited bud (Greene and Autio, 1994). The effectiveness of this technique is attributed to blocking or preventing the downward movement of auxin from the shoot tip. Auxin is the hormone that is primarily responsible for preventing lateral buds from growing. Experience has shown that at least a 2 mm, and, preferably a 4 mm (McArtney and Obermiller, 2015), strip of bark must be removed from immediately above the bud and this bark removal must extend around at least one-third of the stem. Use of a hacksaw blade, either single or double, is a rapid and useful tool to remove this strip of bark in the orchard. Callus will start to form at the cut, and once the callus grows together and fills the gap, auxin again can migrate down to the lateral bud, re-establishing apical dominance thus slowing or stopping the growth of the lateral shoot. Therefore, severing the bark with a knife blade or other sharp instrument that removes a small amount of bark has a temporal influence on stimulating lateral bud growth. The time at which notching is done is important since it influences the number of buds that can be induced to grow. Greene and Autio (1994) reported that notching 'Delicious' buds between 2 and 4 weeks prior to bloom resulted in about 60% of the notched bugs growing, whereas notching done 6 weeks before bloom or at bloom reduced bud break. A larger percentage of the bud broke if buds were located at the

© Burleigh Dodds Science Publishing Limited, 2019. All rights reserved.

tip of a shoot. There was a linear reduction in the percentage of bud that broke the further away the notched bud was from the shoot apex. McArtney and Obermiller (2015) reported that the highest proportion of bud stimulated to grow on 'Granny Smith' apples was when notching was done between bud break and two weeks after bud break. The difference in the optimum time to notch may be related to the early warm temperatures experienced in North Carolina compared with cool temperatures generally experienced in the Northeast US in the spring.

4.2 Branching using PBRs

A number of PBR branching agents have been tested. However, the two primary proprietary products in use today contain 6-BA alone or 6-BA plus GA_{4+7} present in equal amounts. Both products have other uses on apple trees. The 6-BA product has the advantage over the 6-BA plus GA_{4+7} product since the gibberellins in the combined product can inhibit flower bud formation for the next year. Consequently, any tree sprayed with this product may have reduced flower bud formation thus delaying or reducing cropping potential on treated trees the following year (Musacchi and Greene, 2017). On trees where cropping is not planned or not expected for 2 years, a reduction in flower bud formation would not be a problem. 6-BA has little or no direct influence on flower bud formation, although it is used as a proprietary product to thin apples on trees that carry an excessively heavy crop load.

The 6-BA formulation may be applied to young trees in the orchard as either a spray application or painted directly on a bud in latex paint. The latex paint application is made at rates of 5000–7500 mg L^{-1} before bud break. Later application of paint when buds are starting to grow will lead to phytotoxicity and a reduced response. This method is infrequently used. The most common and frequently used method to increase branching on trees is with the use of PBR sprays.

Spray applications of 6-BA or 6-BA + GA_{4+7} when buds start to break at concentrations from 500 to 2500 mg L^{-1} have resulted in variable responses (McArtney and Obermiller, 2015). On similar trees when application is delayed until buds have broken, greater bud break is observed. The variable response is attributed to several factors. In a 2-year study using second leaf 'Granny Smith' and 'Fuji' apples, McArtney and Obermiller (2015) showed that notched buds could cause buds to grow in some instances but if notching was followed by 6-BA application the buds not only started to break but they continued to grow and develop into functional lateral shoots. Apparently, the buds created a somewhat impervious barrier to entrance of the 6-BA spray and notching opened an avenue for the spray to enter and allow the continued growth of the bud.

© Burleigh Dodds Science Publishing Limited, 2019. All rights reserved.

Application of PBRs to stimulate lateral branch development on second and third leaf trees is not a routine management activity. It has been observed in many orchards that using cultural and management techniques that assure good tree growth may be sufficient to encourage the development of an appropriate number of lateral branches, thus not requiring intervention with PBRs.

5 Control of vegetative growth

Management of the growth of apple trees is an ongoing activity. The goal is to maintain a balance between vegetative growth and cropping but management is only partially under the control of the orchardist. Frost, or freeze damage, during the winter may kill flowers thus reducing the crop. Under these circumstances, terminal growth is frequently excessive and requires supplemental control. The use of PGRs is the most effective, convenient and cost-effective tool available to accomplish this and they can be used occasionally or yearly depending on the growth control required. There are two primary growth retardants in general use on apples today: prohexadione-calcium (Pro-Ca) and ethephon (Ethrel).

5.1 Growth control using prohexadione-calcium

Pro-Ca is a gibberellin biosynthesis inhibitor. Because it blocks GA synthesis late in the biosynthetic pathway (between GA_{20} and GA_{1}) (Evans et al., 1999) it has relatively few effects that can't be attributed directly to inhibition of GA synthesis. The primary reason to apply Pro-Ca is to reduce terminal growth by reducing internode length. A consequence of this is that there is less crowding between trees due to shoot competition, light penetration into the tree is increased which then leads to improved fruit colour and fruit quality and improved flower bud formation. The amount of pruning is also reduced.

There are three Pro-Ca formulations available. Apogee® and Kudos® contain 27.5% a.i., whereas Regalis® is a 10% a.i. formulation. Preparation for spraying is different for the 27.5% and 10% a.i. products. Pro-Ca may precipitate out of solution if 'hard' water is used. To remedy this situation when using the 27.5% product you should add an equivalent weight of ammonium sulphate to Apogee® or Kudos® to the spray tank. Alternatively, a commercial water conditioner may be used following label directions. Regalis® is the 10% a.i. formulation that does not require water adjustment since it is formulated with the proper amount of ammonium sulphate. A non-ionic or organosilicate-containing surfactant is recommended to use with all formulations. Pro-Ca should not be applied with any calcium sprays since it will reduce the effectiveness of the Pro-Ca.

© Burleigh Dodds Science Publishing Limited, 2019. All rights reserved.

It requires 10-14 days after application for growth control to become effective (Fig. 6). Regardless of the rate applied, the reduction in growth was initiated at approximately the same time. Apple shoot growth is most rapid in early near bloom. Therefore, to achieve the greatest growth control the initial Pro-Ca application should be made as soon as there is sufficient leaf area to absorb Pro-Ca so that growth control is in place when rapid shoot growth begins. Measureable and commercially significant growth retardation can be achieved with an initial application of 3 oz/100 gal (62.5 mg L^{-1}). Slightly better growth control is possible with rates of 6-12 oz/100 gal (125-250 mg L^{-1}). A higher rate may be selected for the first application which may help compensate for the reduced leaf absorptive area present in the tree. An additional application is required 2-4 weeks after the first and follow-up applications are frequently needed. As the season progresses acceptable growth control can be achieved with lower rates. The amount applied and the number of applications required for season-long control depends on several factors including the cultivar, vigour of the tree, the rootstock, soil, the weather, moisture during the growing season and the geographical location. Usually more applications are required in orchards in warm climates where the growing season is longer and terminal growth continues later into the season. Typically 2-3 applications of Pro-Ca would be used for season-long control in the major fruit growing areas. The highest rate should be applied in the first application and lower rates may be used later.

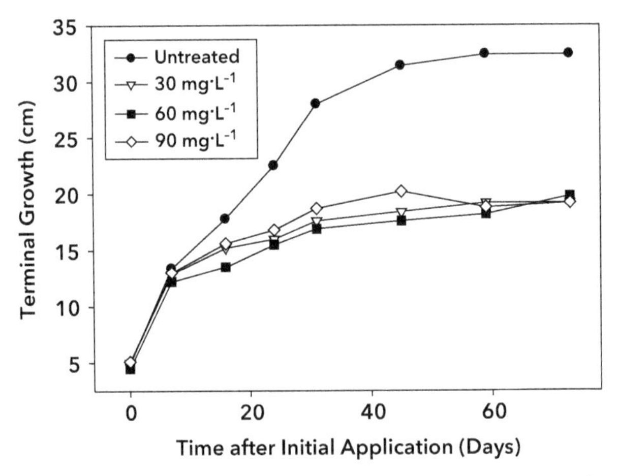

Figure 6 Influence of the rate of prohexadione-calcium (Pro-Ca) applied starting at petal fall on terminal growth of 'Macoun' apples. Regardless of the initial rate of Pro-Ca applied, the time at which growth control was initiated and the extent of growth control was similar. Source: Greene (1999).

© Burleigh Dodds Science Publishing Limited, 2019. All rights reserved.

Pro-Ca often results in increased fruit set (Table 2). It does this by reducing the intensity or extent of 'June drop'. This is the same physiological period that chemical thinners operate in. Pro-Ca inhibits drop of some fruit that would normally be encouraged to drop by post-bloom chemical thinners. Therefore, when Pro-Ca is used in blocks where thinners are applied, it is recommended that more aggressive thinning should be done to achieve an appropriate final fruit set.

Pro-Ca provides excellent control of shoot fire blight. It does not have any antibacterial activity to directly inhibit the fire blight bacteria. It is believed that it acts indirectly by thickening the cell walls in the shoot thus making them less permeable and susceptible to infection. The initial application should be made early, at petal fall of the king flower. Since effective control depends on growth retardation, the spray will only confer resistance to fire blight when growth retardation has been initiated, 10–14 days after application. Rates of 6–12 oz/100 gal (125–250 mg L^{-1}) are generally used. Pro-Ca does not afford any direct protection from blossom blight. Pro-Ca is a very good alternative to the use of antibiotics for control of shoot fire blight bacteria. Since Pro-Ca works by altering the cell structure of the inhibited shoot, bacteria can't build up resistance to shoot fire blight. Antibiotics can be reserved for blossom blight and control of shoot blight following a traumatic event in the orchard such as hail.

5.2 Control of vegetative growth with ethephon

Ethephon is a very effective growth retardant especially when applied from bloom to a week after bloom. It is also a good fruit thinner when applied over the same time period. Consequently, when ethephon is used to control growth it may also cause significant crop reduction. It is generally accepted that ethephon should not be used routinely for growth control on bearing trees due to its thinning potential. Ethephon may be useful in orchards that have lost a crop due to frost. Application of 500 mg L^{-1} when shoots are 10–15 cm in

Table 2 Influence of prohexadione-calcium on fruit set and return bloom when applied at petal fall on 'McIntosh'/M.7 apples to control vegetative growth

Treatment (mg L^{-1})	Fruit per cm limb cross-sectional area	Return bloom per cm limb cross-sectional area
0	5.8	8.4
125	6.2	6.1
250	8.2	4.1
375	9.8	3.3
Significance	l***	l***

Source: Greene (1999).

© Burleigh Dodds Science Publishing Limited, 2019. All rights reserved.

length will retard growth and help the trees to re-establish the delicate balance between vegetative growth and cropping. It should be noted that ethephon promotes flower bud initiation on apples and when applied on bearing trees the potential exists that a 'snowball bloom' may be promoted the following year. This may lead to the trees slipping into a biennial bearing cycle because of the increased difficulty in chemically thinning, especially on difficult to thin cultivars.

Ethephon may be used to control growth of non-bearing trees but this use is far less common now because of the dominant use of precocious dwarfing rootstocks in new plantings. On non-bearing trees, rates of 300–500 mg L^{-1} are commonly applied when the terminal growth reaches 7–12 cm. Growth control is more difficult on trees propagated on vigorous rootstock. Ethephon will enhance flower bud formation on these more vigorous trees.

5.3 Promotion of flowering on bearing trees

Apple trees tend to be biennial, producing a crop one year followed by a year where production is low or there is no crop, due primarily to the reduction in flower bud formation. Cultivars differ in their biennial bearing tendencies. NAA and ethephon are used to thin fruit and they also have the capability to increase flower bud formation, independent of their ability to thin. If these PGRs are applied enough times, at low rates and not during the normal thinning 'window of opportunity' (bloom to 25 mm) they have the ability under some circumstances to increase flower bud formation without causing thinning. There is a general consensus, based on experience, that either 5 mg L^{-1} NAA or 100 mg L^{-1} ethephon may be effective when fruit reach 25–30 mm in diameter. The spray should be repeated three or four more times at 7–10 day intervals. The effectiveness of these sprays differs from year to year and there is no guarantee that they will actually increase flowering. However, this is the best current option available to influence flowering on bearing trees once a crop has been set. Ethephon should not be used on 'McIntosh', 'Macoun' or 'Honeycrisp' or other 'McIntosh' types because if applied there is the possibility of advancing ripening if ethephon is used even at this late timing.

6 Crop load management

6.1 Chemical thinning

Chemical thinning has been practised in orchards since it was first recognized to help alleviate biennial bearing in the 1930s. Initial efforts were made to learn how to most effectively use the thinners available. Many empirical thinning studies were done but variability and lack of consistency of the thinning responses persisted. More recently, progress has been made and more emphasis has

© Burleigh Dodds Science Publishing Limited, 2019. All rights reserved.

been placed on attempting to determine how individual thinners work, gaining a greater understanding of how environmental conditions influence fruit set and thinner response and recognizing the changing susceptibility of fruit to thinners at different physiological stages of fruit development. Interest in thinning has continued to increase because of the diminishing pool of workers available to hand thin orchards and because of the economic consequences fruit growers suffer from reduced crop value due to insufficient thinning.

A number of studies have been done to determine the mode of action of specific thinners. While there is no consensus on a specific mode of action, it appears that they may act by either influencing the amount of carbohydrate available to developing fruit or they may influence the auxin production or movement from the fruit. Bangerth (2004) reported that young fruit that persist to harvest have a continuous supply of auxin moving from the developing fruit, down the pedicels and through the abscission zone. Abscising fruit have a reduced supply of auxin moving from the fruit through the abscission layer which then triggers the production of enzymes that result in the destruction of these cells in the abscission zone leading to fruit abscission. When carbon supply available to young developing fruit drops to a critical low level, abscission of fruitlets is irreversibly initiated. When the carbohydrate available to young developing fruit is reduced either by shading (Byers et al., 1990a) or by applying photosynthetic inhibitors (Byers et al., 1990b), fruit could be induced to abscise. Lakso et al. (2006) showed that conditions that result in low availability of carbohydrates during peak carbohydrate demand periods result in fruit abscission. It may be that the carbohydrate deficit may serve as the triggering event which then signals the reduction and movement of auxins through the abscission zone.

Variability of the weather has been recognized as a major contributing factor in the lack of thinning consistency, even in the same block using the same thinners and the same rates in different years. Many researchers have made contributions to help quantify and clarify the weather influence on thinning. Williams and Edgerton (1981) identified 15 factors that could influence thinning response but temperature and solar radiation before, during and the 3-4 days following thinner application have emerged as the factors that appear to have the greatest influence on the response to thinners.

Apple may be chemically thinned over a wide range of times starting as early as bloom and extending until fruit reach about the 25 mm fruit growth stage. However, as a fruit grows it undergoes physiological changes that of necessity require use of different thinning strategies and different thinning chemicals. During bloom the objective of applied thinners is to damage the stigma or interfere with pollen germination, pollen tube growth or fertilization. Once the king flower has been successfully pollinated and the pollen tube has grown down to the ovary most commonly used blossom thinners inflict some

© Burleigh Dodds Science Publishing Limited, 2019. All rights reserved.

type of damage to the remaining unpollinated flowers. Blossom thinners are routinely used in arid growing regions. If they are applied in humid growing climates, these compounds may cause some phytotoxicity, which in some cases may be severe. In addition to potential damage, growers in these humid growing regions appear to be reluctant to thin before the pollination period ends and when the danger of frost has passed so that they can assess the potential for initial fruit set. Thinning decisions in humid regions should be re-evaluated since thinning at bloom provides the best opportunity to increase fruit size and ensure return bloom (Batjer and Hoffman, 1951) and the potential danger of thinning at bloom may be more perceived than real or that can actually be documented.

Petal fall, immediately after bloom, affords an extremely important thinning opportunity. The bloom period has passed and the danger of frost has diminished. Meaningful thinning can occur but over thinning rarely occurs if thinners are applied at this time (Greene, 2002). Physiologically, cell division is occurring and seed development is starting but in the receptacle (fruit) visible changes are small and subtle and difficult to measure.

Fruit are most vulnerable to applied thinners during the 7–14 mm growth stage when typically there are 3-5 fruit present on each spur and they are in the log growth phase. Storage carbohydrate has been nearly depleted and the primary source of carbohydrate for growing fruit is now supplied by the leaves associated with each spur. Impact of photosynthate from other portions of the tree does not generally occur until later when shoot growth reaches 10 inches (25 cm). Chemical thinners applied at this time create a stress which may amplify the differences between photosynthate required for all fruit to persist and the ability of the spurs to supply carbohydrate. When a carbohydrate deficit occurs, an internal signal is given resulting in the abscission of the weakest fruit.

As fruit grow larger (>15 mm) they become progressively more difficult to thin (Greene, 2002; Schwallier, 1996). At this development stage, ethylene production declines, there is less intra-fruit competition due to previous fruit abscission within the spur and fruit are starting to accumulate starch. Carbohydrate deficits are less likely to occur unless unusual weather conditions occur. As fruit approach 17–20 mm diameter they regain more sensitivity to ethylene and ethephon or 1-amino-cyclopropane-1-carboxylic acid (ACC) application appear to be more effective. Once fruit have grown to about 25 mm, further reduction in crop load is generally very difficult.

6.2 Thinning chemicals

Thinning chemicals may be roughly placed in two categories: caustic thinners and hormone thinners. Caustic thinners are generally applied during bloom with the objective of damaging the flower, flower parts, or damaging or

© Burleigh Dodds Science Publishing Limited, 2019. All rights reserved.

preventing the pollen germination or inhibiting pollen tube growth and thus preventing fertilization of the ovule. The king flower is generally the largest flower in the cluster and the first to open so the goal is to assure that flower set fruit while inhibiting set of three remaining flowers. Attempts are made to time the blossom thinning application to allow pollination of the king then apply a blossom thinner to inhibit set of the remaining flowers in the cluster. Sometimes the window of opportunity to apply blossom thinners is narrow and can be measured in hours.

A concerted efforts was made to find a replacement for the popular blossom thinner Elgetol (DNOC, sodium 4, 6-dinitro-*ortho*-cresylate) when its registration was cancelled in 1990, but few replacement thinners are in general use or remain registered for this purpose. Lime sulphur is a very commonly used blossom thinner. It is frequently combined with fish oil or other oils to enhance activity. ATS is a foliar nitrogen fertilizer. Although not specifically registered for use as a thinner, if applied to trees at bloom at a rate of 2–4 gal/100 gal (20–40 L/1000 L) it has been shown to be a viable thinner.

Blossom thinners are routinely used in dry, low humidity climates. If blossom thinners are applied before wet or high humidity periods it is not uncommon to have a significant amount of foliage damage. Spur leaves are very important to support fruit cell division and early fruit growth. The observation has been made that application of blossom thinners in humid-wet conditions may lead to thinning but it may not be manifest in increased fruit size at harvest due to foliage damage. Blossom thinning removes fruit early, thus providing the best opportunity to increase return bloom. It therefore remains a viable option to be used, especially on cultivars that are noted for biennial bearing and in areas where weather does not routinely create a carbohydrate stress within the tree during the 7–14 mm fruit size period.

The majority of thinning is done using hormone-type thinners. Unlike caustic thinners that act by causing some type of damage, hormonal thinners act by creating some form of stress and interacting with some physiological events in the plant, which are frequently modified by the weather. Fruit physiological changes occur as the fruit continues to grow. Therefore, selection of thinner(s) may change over time as fruit grows and undergoes developmental changes. The mode of action of specific thinners is not definitively known, although many investigators have identified physiological events that may be affected such as a reduction in photosynthesis, an increase in respiration, limitation of photosynthate transport or enhanced production of ethylene (Dennis, 2002). As a fruit increases in diameter above 15 mm it becomes progressively more difficult for thinners to cause fruit abscission. In general, when fruit diameter exceeds 25 mm it no longer responds to hormonal thinners (Greene, 2002).

The time applied and the circumstances under which individual thinners are used has been previously reviewed: Byers (2003), Greene (2002), Greene

© Burleigh Dodds Science Publishing Limited, 2019. All rights reserved.

and Costa (2013), Schwallier (1996), Wertheim (2000) and Williams and Edgerton (1981).

6.2.1 Naphthaleneacetic acid (NAA)

NAA is the oldest and perhaps the most potent of the thinners now in general use. It may be used in concentrations from 3 to 20 mg L^{-1} but the rates used are generally between 5 and 12 mg L^{-1}. It may be applied between bloom and about 15 mm fruit size. When higher rates are applied on fruit larger than 15 mm especially in very warm weather, reduced fruit size or no increase in fruit size may be noted even though crop load is reduced. It is frequently chosen as the thinner to use on very difficult to thin cultivars. Because NAA has the reputation for being a strong thinner, growers use it with caution. It is commonly combined at lower rates with other thinners, especially carbaryl.

6.2.2 Naphthaleneacetamide (NAAm, NAD)

NAD was tested and evaluated during the same time that NAA was evaluated. NAA became the thinner of choice because it was more potent. It is generally used at a rate of 40–50 mg L^{-1} and the time of application is generally at bloom or petal fall. It can cause pygmy fruit formation when applied late and on 'Delicious' (and perhaps other cultivars). It can cause leaf epinasty but it is far less severe than that caused by NAA. NAD is frequently used in combination with carbaryl and then followed by another thinner at the 7–14 mm stage.

6.2.3 Carbaryl

Carbaryl is perhaps the most versatile thinner in general use. It is an insecticide and a mild thinner that can be used over a wide range of developmental stages from petal fall until fruit reach 18–20 mm. It is a popular thinner since it rarely over thins. Unfortunately, it is very toxic to bees so the earliest time it should be used is after bloom, when beehives have been removed from the orchard. Unlike most thinners, the thinning is concentration independent.

6.2.4 Benzyladenine (BA)

BA is the newest thinner to be registered for use on apples. Effective thinning concentrations are 50–150 mg L^{-1}, although the most commonly used rates are between 75 and 100 mg L^{-1}. BA when used alone is a mild thinner but when combined with carbaryl it is considered a potent thinner that may be similar in strength to NAA. BA can increase fruit size beyond that attributed to a reduction in crop load due to its ability to increase cell division (Wismer et al., 1995). It is

© Burleigh Dodds Science Publishing Limited, 2019. All rights reserved.

most effective when temperatures in the 3–4 days following application rise to 22–25°C.

6.2.5 Ethephon

Ethephon is generally not considered a mainstream thinner in most growing regions. Flower and fruit susceptibility to ethephon varies depending on the stage of development. Ethephon is an effective blossom thinner (Jones et al., 1990) but it loses some of its effectiveness as fruit develops. Fruit redevelop thinning sensitivity between 16 and 22 mm (Marini, 1996). It is the thinner that most effectively thins larger fruit (18–25 mm), although thinning may be inconsistent when applied during this period. It is sometimes referred to as a 'last chance thinner'. The primary use of ethephon now appears to be in orchards where growers failed to achieve acceptable thinning during the normal thinning time period.

6.2.6 Metamitron (Brevis®)

The newest thinner, to be evaluated and in some countries registered for use as a thinner, is metamitron (McArtney and Obermiller, 2012). It was originally registered for use as an herbicide to be used on sugar beets. It is a photosynthetic inhibitor that blocks electron transfer in photosystem II (Abbaspoor et al., 2006). When used at rates of 300–400 mg L^{-1} it thins apples, presumably by reducing the carbon balance within a tree resulting in reduced fruit set. It appears to have the greatest activity when applied during the 7–14 mm stage when apples are generally most sensitive to a reduction in photosynthesis and when the weather conditions following application indicate that the tree will be under a carbon deficit. Experience has shown that having prevailing weather conditions 4–5 days following application that create a negative carbon balance are necessary for metamitron to thin effectively. When applied at rates above 300 mg L^{-1} some minor leaf chlorosis may be possible, but it does not appear to be commercially important.

6.3 Precision thinning

Thinning is the most important management activity a grower is required to do. Insufficient thinning will result in the harvest of small, poor quality fruit and reduced return bloom which leads to reduced income for 2 years. Establishment and growing costs have escalated to a point where large fluctuations in income are no longer economically sustainable. Precision thinning is a strategy developed to assure that consistent thinning is achieved each year thus assuring a somewhat steady income. When fully implemented, Precision Crop

© Burleigh Dodds Science Publishing Limited, 2019. All rights reserved.

Load Management involves three steps. These include (1) precision dormant pruning adjustment, use of computer models to (2) predict thinning prior to thinner application and (3) early prediction of effectiveness of the thinning sprays (Robinson et al., 2016).

6.3.1 Precision dormant thinning

Precision dormant thinning will be initiated by growers who determine during the dormant season the number of fruit that they would like to harvest from a tree or unit. This is a deviation from earlier orchard practices where crop loads per tree were generally not preset. Setting fruit crop goals is quite easy on new high-density plantings where tree numbers and total desired yield per acre are known. It is much more difficult on larger trees where branch subunits are relied on.

Once the target number of fruit per tree is established, the number of flower buds on a tree needs to be determined. This may be challenging at first but with a little practice it can be done somewhat accurately, even during the dormant season. Select five uniform and representative trees in a block and count all flower buds. Next, calculate the blossom cluster averages for the measured trees. Multiply the predetermined and desired number of fruit on each tree by 1.5 and this will for most cultivars be the number of flower buds that should remain after pruning.

Assume an orchard is planted at a 12 × 3 ft. (0.9 × 3.7 m) spacing it would contain 1210 trees (2990 trees ha^{-1}). If the desired yield is 1200 bu $acre^{-1}$ (59 300 kg ha^{-1}) then 100 fruit per tree would be the desired final crop load. If you multiply 100 fruit by 1.5 the target number of flower clusters to remain should be 150. Trees would be pruned in an attempt to reduce the flower cluster number down to the calculated number. Initially, pruning followed by counting would be required to fine-tune the process.

Thinning at pruning time offers several advantages. It reduces early competition among flowers. Weaker flowers may be pruned off thus allowing resources and new growth on a tree to grow the large flower buds that have a greater potential to produce large fruit.

6.3.2 Apple carbon balance model

Precision thinning done during the bloom and post-bloom period utilizes two recently developed computer models, the apple carbon balance model and the apple fruit growth model. These two models work together to allow methodical reduction in crop load.

The apple tree carbon balance model was developed by Alan Lakso at Cornell using a mathematical model to estimate apple tree photosynthesis,

© Burleigh Dodds Science Publishing Limited, 2019. All rights reserved.

respiration and growth of fruit, leaves, roots and woody structure (Lakso et al., 2006, 2007). The model uses daily maximum and minimum temperatures and solar radiation to calculate the daily production of carbohydrate and allocates the available carbohydrate to the organs in the tree including developing fruit. Temperature and light are primarily responsible for determining the carbon balance in a tree. In low light intensity and high temperature conditions there would be a carbon deficit and thinning would be easy. Conversely, high light and lower temperatures favour carbohydrate accumulation and thinning would be difficult. The carbon balance model is able to quantify the carbon balance using these inputs. The data can then be used as a guide to determine if thinning at a given time is appropriate and if so it provides guidance about the strength of the thinner to use.

6.3.3 Apple fruit growth model

This model is based on the observation that fruit that abscise following thinner application slow growth well in advance of the time that they actually abscise (Greene et al., 2013). Generally 3-4 days is required for a thinner to affect fruit growth. In this model, spurs are tagged prior to thinner application. At 3-4 days after thinner application all fruit on the tagged spurs are measured. Generally, at 3-4 days after application of the thinner, fruit that are induced to abscise will start to show a reduction in growth rate. Fruit must be at least 6 mm in diameter to start measuring. A second measurement is made 4-5 days after the initial measurement (Fig. 7). At 7-8 days after thinner application a significant

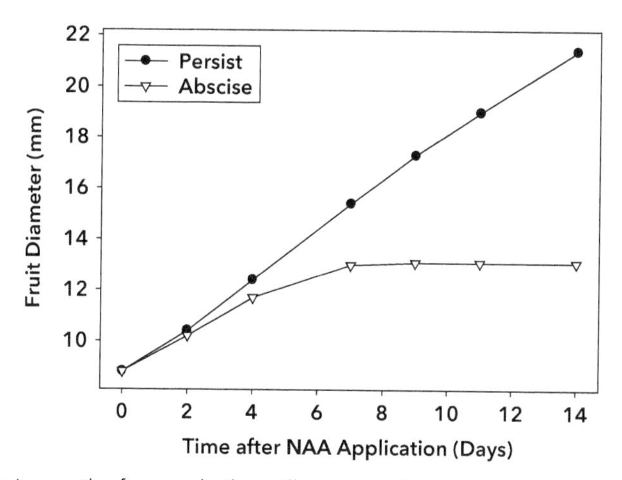

Figure 7 Fruit growth of an apple that will persist to harvest and one that was induced to abscise as a result of a naphthaleneacetic acid (NAA) thinning spray. Fruit that are induced to abscise exhibit slow growth 3-4 days after application and by 8 days after application fruit growth has essentially stopped.

© Burleigh Dodds Science Publishing Limited, 2019. All rights reserved.

reduction in fruit growth can be measured. These data are entered into an Excel spreadsheet programme and the growth rate of each fruit is calculated. Fruit growth rate is a good indication of the fate of young fruit (Fig. 8). This figure illustrates that slow growing fruit will abscise and fast growing fruit will persist. The fruit growth model considers fruit that grow at less than 50% of the rate of the fastest growing fruit will abscise whereas those growing faster than 50% will persist. This model requires establishing a desired end crop load. The spreadsheet predicts the crop load if no further thinning is done. If the predicted crop load is insufficient then a follow-up thinner spray is applied. Fruit will then be measured again to determine if the crop load has been lowered to the target level. Implementation and details for grower use are outlined by Robinson et al. (2013, 2014) and Schwallier and Irish-Brown (2015).

6.3.4 Pollen tube growth model

Blossom thinning is done routinely in areas that are typically arid. It has the advantage over post-bloom thinning since it can have the greatest effect on fruit size and return bloom (Greene, 2002). Thinners act by preventing pollination and fertilization by damaging the anthers, stigma and style of flowers or inhibiting pollen tube growth (McArtney et al., 2006). Blossom thinners can be very erratic since timing is critical for its success. Ideally, there should be a sufficient number of king flowers open and pollinated to achieve a full crop. Once this is accomplished a blossom thinner should be applied to prevent pollination of the remaining flowers. Timing of the spray is important to the success of blossom thinners. The pollen tube growth model

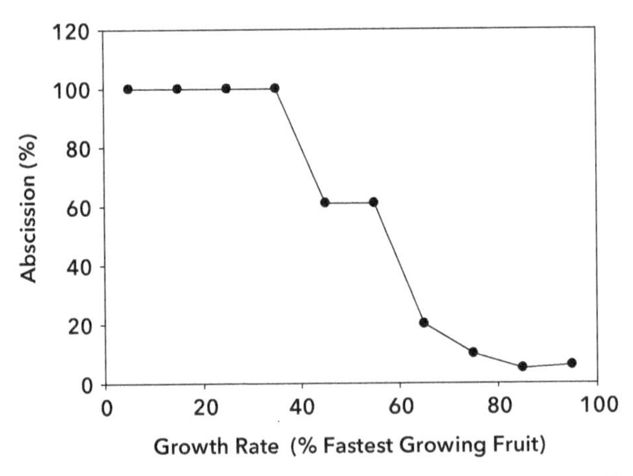

Figure 8 The fruit growth rate 4–8 days after thinner application is an excellent indicator/predictor of which fruit will persist and which will abscise. Source: Greene et al. (2013).

© Burleigh Dodds Science Publishing Limited, 2019. All rights reserved.

was developed to provide guidance to accurately time the application of the blossom thinners (Peck et al., 2016; Yoder et al., 2013). Temperature influences the rate of pollen tube growth. The length of the style is initially measured, and taking temperature into account, the model indicates the time required for the pollen tube to grow through the style to fertilize the ovule. As soon as the model predicts the pollen tube has reached the ovules in the ovary, the blossom thinner should be applied to discourage pollination and fertilization of all lateral flowers. Since pollen tubes grow at variable rates of different cultivars, a model for each cultivar must be established.

7 Influencing flowering and fruit set

Developing and executing a management strategy to assure adequate bloom each year is a significant challenge. A key component to achieve this goal is to successfully manage a chemical thinning programme. However, in situations where less precocious rootstocks are used, PBR treatments may be useful to help control vegetative growth and increase flowering bud formation. The rapid shift to the use of dwarfing rootstock that is occurring has changed the way orchardists approach the problem of regulating flowering and fruiting in an orchard. The most commonly used rootstocks now are very precocious, and with some varieties, excessive flowering and fruit set may occur before the tree has had the opportunity to fill its allotted space. In these situations the major problem is to establish an appropriate balance between allowing fruit set to occur on young trees while limiting the crop load to allow adequate return bloom and growth of young trees. The need for increased flower bud formation in orchards has changed over time. The need remains but the methods used are different.

7.1 Promotion of flowering on non-bearing trees

The age that flower bud formation and fruit set occur on young apple trees is frequently delayed when trees are propagated on less precocious semi-dwarfing rootstocks. This delay in age at which flowering starts limits early production and extends the amount of time required to achieve full return on the investment made in an orchard. There are cultural methods that growers can use to promote flower bud formation. Scoring or girdling are methods of severing the phloem thus reducing or preventing the movement of photosynthate from the top of the tree down to the root system. This starving of the root system encourages flower bud formation while reducing vegetative growth. Scoring generally is the least aggressive method since a knife is often used to make the cut around the trunk. Girdling is more aggressive since it involves the removal of a band of bark on the trunk of

© Burleigh Dodds Science Publishing Limited, 2019. All rights reserved.

a tree. The wider the cut or the larger amount of bark removed the more severe the girdling effect will be. A hacksaw blade may be used to remove a narrow strip whereas a chainsaw can remove a much wider strip of bark. The timing that girdling is generally done is shortly after bloom when shoots are 5–10 mm in length. Even on trees where the use of this method is appropriate, growers are frequently reluctant to use this method of growth control now because it creates an avenue for the fire blight bacteria to enter the tree at a time of year when this devastating disease is frequently active. This is a greater concern now that fire blight has become a more prevalent and widespread problem.

Branch spreading and bending are cultural practices that have been used on trees to reduce vegetative growth and increase flower bud formation. In the past, wooden limb spreaders with sharpened nails in the ends were used on permanent and semi-permanent scaffold branches. This was a very common practice used on trees on vigorous semi-dwarfing rootstock. Many methods have been used to spread limbs and flatten branches on a tree including toothpicks, numerous types of weights and clothespins to mention just a few. More recently branch bending has been used as a method to control growth and increase flower bud formation on large limbs on newly planted trees in a tall spindle planting (Robinson et al., 2008).

The primary PBR that is used to enhance flowering on non-bearing trees is ethephon, although the frequency of use has diminished as the number of trees propagated on semi-dwarfing rootstocks has declined. In general, ethephon is initially applied on trees that are large enough to bear at least a partial crop of apples and this is usually on trees 2–4 years old. The recommended rate is 300–450 mg L^{-1} and the application is usually made 10–14 days after bloom. A reduction in both terminal and lateral growth can be expected. Sometimes application of ethephon at this time does not result in increased bloom or subsequent fruit set and application for a second year may be appropriate (Greene and Lord, 1978).

7.2 Promotion of flowering on bearing trees

Flowering on bearing trees is regulated primarily by crop load adjustment using chemical thinners. However there are several reasons why orchardists may wish to employ supplemental methods to encourage flower bud formation. First, as many trees start to come into production they may be very easily nudged into biennial bearing when 3 or 4 years old if crop load on these young trees is not regulated carefully. Some popular varieties have a natural tendency to become biennial. Chemical thinning is not a precise science since success is often dependent upon the weather conditions working in conjunction with thinners. It is not always possible to time thinner applications with favourable

© Burleigh Dodds Science Publishing Limited, 2019. All rights reserved.

weather since weather forecasts are subject to change after a thinner has been applied. In some years favourable weather may never occur. Where conditions are not favourable, trees may be overset at the end of 'June drop'. There is a period of time after the end of June drop that flowering can be affected by PBRs, although thinning may not be possible. There are two PBRs that may be used on bearing trees. NAA is the most common PBR used on bearing trees for this purpose. NAA at 5-7.5 mg L^{-1} can be applied 3-4 times at 7-10 day intervals, starting 4-6 weeks after bloom when fruit size is generally greater than 25 mm. Alternatively, ethephon at 100-150 mg L^{-1} may be used in a similar manner to NAA (Byers, 1993). These treatments do not guarantee increased flower bud formation, especially on difficult varieties to increase bloom on such as 'Honeycrisp', but they are the best 'insurance policy' available to affect bloom on bearing trees.

7.3 Inhibition of flowering

Regulation of flower bud formation in apples is generally accomplished by causing fruit abscission during the first 3 weeks after bloom. Fruit or seeds in fruit are primarily responsible for the inhibition of flower bud formation in spurs (Chan and Cain, 1967). It is widely acknowledged that gibberellins produced in the seeds of developing fruit diffuse out of the seed, migrate to the bourse bud associated with that spur and inhibit flower bud formation in the spur. By simply removing the fruit from some spurs, flower bud formation on those spurs is then allowed to proceed. Therefore, the removal of developing fruit does not increase flower bud formation per se, but rather removal of fruit removes the flower bud inhibitor, gibberellin present in the seeds, allowing flower bud formation to proceed unimpeded in the bourse bud associated with the spur.

Apples and pears have a biennial bearing habit which results in a heavy bloom and fruit set in the 'on year', followed by a year where there are few flowers present which result in a few fruit setting, the 'off year'. The normal way to break or modify this cycle is to thin trees in the 'on year' which will then leave spurs with no fruit allowing for flower bud formation to proceed. However, some cultivars are difficult to thin making this strategy difficult and trees such as peach can't be thinned with hormonal sprays effective on apples and pears. It is possible to apply gibberellins to trees in the 'off year' and even out flowering in the 'on year' by inhibiting flower bud formation. To be most effective the gibberellins must be applied early in the season to achieve maximum effect, although some inhibition of flowering can occur up to 2 months after bloom (Tromp, 1982). McArtney (1994) reported that a single spray of gibberellin A_3 (GA_3) or GA_{4+7} at full bloom in the 'off year' reduced the subsequent season's flowering in the 'on year' and the severity of biennial bearing in 'Braeburn' apple.

© Burleigh Dodds Science Publishing Limited, 2019. All rights reserved.

7.4 Commercial PBRs that inhibit flower bud formation

There are proprietary gibberellin products that are registered for use on apples, although inhibition of flowering is not one of the registered label uses. GA_{4+7} (Provide®, Novagib®, TypRus®) is used to reduce fruit russetting and to reduce cracking on some apple varieties. Although applications for russet control start at petal fall, inhibition of flowering probably does not occur as the result of this treatment due primarily to the low rate used. A 50% 6-BA plus 50% GA_{4+7} formulation (Promalin®, Perlan®, Typy®) has several uses in tree fruit but the use that has the greatest influence on flowering is when employed to increase branching on trees in the nursery and to enhance branching on newly planted trees. Application is made early in the spring. The high rates used at this time can inhibit flower bud formation. This may be considered a positive response since it is generally undesirable to have fruit development too early on young trees which in turn could delay tree development and filling of the allotted space.

7.5 Increase fruit set

In tree fruit production emphasis is frequently placed on strategies to reduce fruit set as a way to improve fruit quality at harvest and to encourage good return bloom. However, there are a number of situations where it is appropriate to increase fruit set. Situations where increased fruit would be appropriate include frost damage to flowers, poor weather during the pollination period, light blooming trees, winter cold damage to buds and on cultivars where natural fruit set is typically low.

Foliar sprays of gibberellins can increase fruit set and parthenocarpic fruit development on pears when applied near bloom and this strategy can be used commercially to increase cropping on blocks of pear trees that have poor natural fruit set. Yarushnykov and Blanke (2005) reported that gibberellins could increase fruit set of pears after frost damage at bloom. Gibberellins, particularly GA_{4+7} can cause parthenocarpic fruit set on apples (Bukovac, 1963) but the response is quite variety dependent. McArtney et al. (2014) reported that the proprietary mixture of 6-BA plus GA_{4+7} increased parthenocarpic fruit set on frost damaged apple trees. While final fruit set was reduced in all trees by the frost, treated trees had a significantly higher yield than non-treated trees. It appears that the treatments must be applied within 4 days of the frost event to be effective.

It was established as early as 1980 that AVG could increase the fruit set on apples (Williams, 1980). The commercial formulation of AVG (ReTain®) was not registered for commercial use until 1997 and addition of increasing fruit set on the label did not occur until later. The suggested time of application is from the pink stage until full bloom. Use for this purpose is not widespread, in part because the need is frequently not great and application at bloom reduces

© Burleigh Dodds Science Publishing Limited, 2019. All rights reserved.

the amount that can be applied nearer harvest where the use is common and widespread. Some cherry cultivars (e.g. Regina) routinely have low fruit set and AVG application between the balloon stage and first bloom can increase set (Bound et al., 2014).

Early in the development of prohexadione-calcium (Pro-Ca) it was recognized that it had the potential to increase the fruit set of apples when applied to control vegetative growth. Application timing is from pink to when the new shoot growth is 2–5 cm in length and it requires 10–14 days for the compound to start to exhibit signs of growth retardation. This time coincides somewhat with the initiation of 'June drop'. Pro-Ca generally does not increase fruit set per se but it may inhibit June drop which frequently does lead to an increase in final fruit set. Pro-Ca can negate some of the abscission-promoting effects of chemical thinners. Therefore, it is routinely suggested for orchardists to be more aggressive in their thinning programmes on blocks treated with Pro-Ca to help negate the increased fruit set tendencies induced by Pro-Ca. However, the primary reason to apply Pro-Ca remains to achieve growth retardation and to aid in the control of shoot fire blight rather than using it to specifically increase fruit set.

8 Pre-harvest application of plant bioregulators

PBRs are frequently used during the 4 weeks prior to normal harvest to influence the ripening process and to retard pre-harvest drop. The PBRs used and the time of application will influence the time of ripening, fruit quality, storage potential and fruit abscission.

8.1 Delay pre-harvest drop and retard ripening

The apple is a climacteric fruit and as such the onset of the climacteric signals a definitive stage in the development of maturation and ripening (Watkins, 2003). In climacteric fruit, ethylene advances the time of the onset of the climacteric. Ethylene is an endogenous hormone that plays a dominant role in the ripening process in apples and an initiator of pre-harvest drop. Stressful conditions (drought, high temperature) can trigger and exacerbate abscission prior to fruit reaching an acceptable level of maturity for harvest. Additionally, if orchardists do not have sufficient labour to harvest fruit in a timely manner, drop control measures would be required to avoid substantial economic loss especially on drop-prone cultivars.

Many commercially important apple cultivars suffer from the pre-harvest drop malady before they have reached maturity that would indicate optimum commercial harvest date (Greene et al., 2014). Drop-prone cultivars such as 'McIntosh', 'Macoun' and 'Honeycrisp' may suffer substantial loss due to drop.

© Burleigh Dodds Science Publishing Limited, 2019. All rights reserved.

With these cultivars, it is often normal to have 20% of the fruit drop prior to harvest and in some years, it is not uncommon to have losses up to 50%.

8.1.1 Pre-harvest drop control PBRs

There are three commercially available drop control compounds: NAA (many formulations), AVG (ReTain®) and 1-methylcyclopropene (1-MCP, Harvista™).

8.1.1.1 Aminoethoxyvinylglycine

AVG is a naturally occurring PBR that was discovered in the early 1970s. Its primary mode of action is to inhibit ethylene biosynthesis (Boller et al., 1979). Bangerth (1978) was the first to demonstrate that it could inhibit pre-harvest drop of apples. Development of this compound for drop control and other uses was not continued because of the high cost of production. In response to the withdrawal of registration of daminozide in 1989, development of AVG was resumed. Abbott Laboratories registered a proprietary formulation of AVG in 1997, ReTain®, to control pre-harvest drop and delay ripening of apples.

The extent of and the strategy for pre-harvest drop control is affected by the cultivar, the time of application, the objective of the application and the amount applied (Greene and Schupp, 2004; Schupp and Greene, 2004). For most cultivars where pre-harvest drop is a significant problem, the initial application is generally made 3–4 weeks before the anticipated start of normal harvest. Following application, it generally requires 10–14 days for drop control to become effective (Greene, 2006). If an application is made later, there is a chance that early drop may occur before AVG can become effective. Earlier application may result in early drop control but loss of drop control later during the harvest season when pre-harvest drop control is characteristically more severe. Split applications are a partial solution.

For most drop-prone cultivars, an initial application of one pouch (333 g of 15% ReTain® formulation per acre, 825 g ha^{-1}) is used. This rate will frequently control and restrict drop loss to about 20% between 5 and 6 weeks after application. It has long been known that higher rates of AVG will increase drop control. In 2015, the ReTain® label was updated to allow application of two 333 g packages of ReTain® per acre (1650 g ha^{-1}) in a season. This additional AVG further extended the effective period of drop control (Fig. 9). A lower (half) rate is usually used on low ethylene producing cultivars such as 'Gala' and 'Honeycrisp'. While 'Gala' does not suffer from pre-harvest drop, ReTain® is used to delay ripening which helps reduce fruit cracking and developing an oily 'greasy' feel which is frequently associated with ripe fruit. If full rates are used on these cultivars, ripening and red colour may be delayed too

© Burleigh Dodds Science Publishing Limited, 2019. All rights reserved.

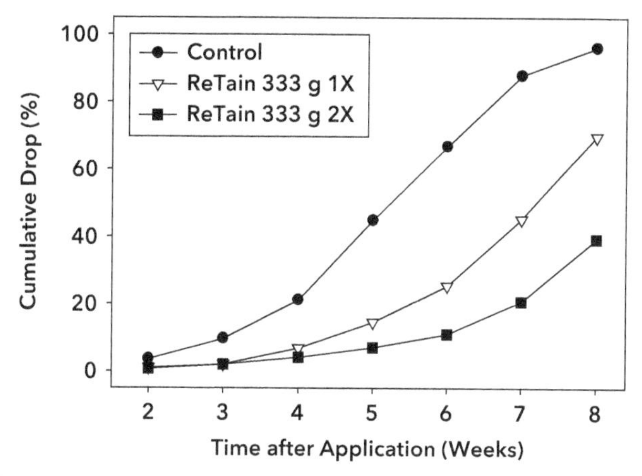

Figure 9 Influence of one full rate application of ReTain® (333 g $acre^{-1}$, 825 g ha^{-1}) or two full rates on pre-harvest drop of 'McIntosh' apples. The double application slows drop and allows greater drop control for a longer period of time. If one uses 20% drop as an acceptable loss, then the double application extends the period of drop control for 7-10 additional days.

long (3-4 weeks). The maximum amount of ReTain® that can be applied in an orchard is two pouches (666 g $acre^{-1}$ (1659 g ha^{-1} $year^{-1}$)). This amount may be applied at one time or it can be applied in two or more instalments. The timing and the amounts applied at each time depend on the objectives of the grower. The earlier in the season ReTain® is applied, the greater the delay in ripening. Maximum delay in ripening would occur if the highest rate of ReTain® was applied 4 weeks before harvest. Frequently growers use a split application strategy to have less reduction in ripening and use later applications to extend the period of drop control.

A 100% organosilicone surfactant accompanying ReTain® will enhance foliar uptake and improve performance but not all surfactants are equally effective. Currently, Silwet L-77 and Sylguard 309 are recommended for the United States at the rate of 0.1% v/v. Other organosilicate surfactants may be effective, but they would require testing. The organosilicone surfactants also provide some rain fastness if the spray has had a sufficient amount of time to dry (Greene, 2006).

8.1.1.2 1-methylcyclopropene (1-MCP, Harvista™)

1-MCP is an inhibitor of ethylene action that acts by binding irreversibly to ethylene receptors in the treated tissue. If binding sites are occupied by 1-MCP a plant is unable to respond to ethylene, even if present (Fan et al., 1999). It is applied to harvested fruit in sealed rooms. Because 1-MCP is a

© Burleigh Dodds Science Publishing Limited, 2019. All rights reserved.

gas, application to trees in the orchard provides logistical challenges. An initial approach was to mix 1-MCP with oil and water immediately prior to application to ensure a more targeted application. Harvista™ can be applied using a proprietary in-line injections system developed by AgroFresh. However, AgroFresh is developing its own sprayer adaptation to allow the application of Harvista™.

The pre-harvest drop that occurs on trees treated with ReTain® differs from that on trees treated with Harvista® (Fig. 10). On ReTain®-treated trees drop occurs slowly at first and then after 5-6 weeks the rate of drop accelerates as more fruit become climacteric. On Harvista™-treated trees, initial drop control is nearly complete but new ethylene binding sites are generated in the fruit during the ripening process. Ethylene produced can then attach to those new binding sites, thus initiating the abscission process. This results in a rather rapid increase in pre-harvest drop that can be substantial.

Harvista™ like ReTain® may also delay red colour development, starch degradation, flesh firmness loss, degradation of starch and water core development. Following application, drop control may be achieved within 3-5 days. Since it can be applied closer to normal harvest, fruit have a longer time to mature on the tree normally and to develop red colour compared with ReTain® which must be applied earlier. Loss of drop control with Harvista™ is associated with ripening fruit starting to generate new ethylene binding sites which can then respond to ethylene being produced by the fruit. A follow-up application of Harvista™ may saturate newly generated binding sites thus significantly extending the period of drop control (Fig. 11).

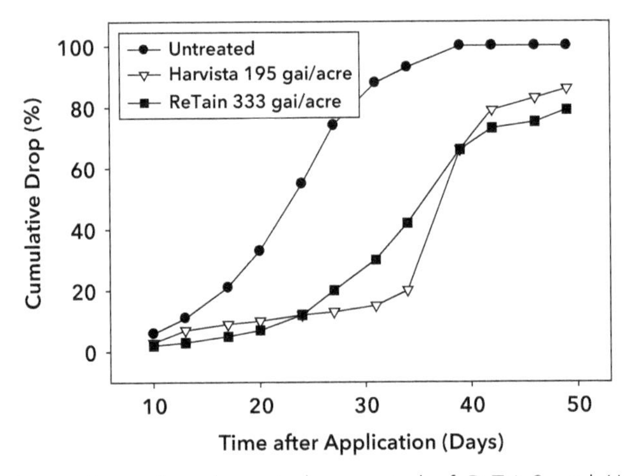

Figure 10 Comparison of pre-harvest drop control of ReTain® and Harvista™ on 'McIntosh' apples. Both compounds retarded drop comparably early in the season. Drop control on Harvista™-treated trees was lost rapidly but drop from ReTain®-treated trees was more gradual.

© Burleigh Dodds Science Publishing Limited, 2019. All rights reserved.

8.1.1.3 Naphthaleneacetic acid (NAA)

It was discovered in the 1930s that auxins can inhibit pre-harvest drop of apples. Over the past 75 years several auxins have been registered for pre-harvest drop control, but NAA remains the only auxin registered for use for this purpose.

The standard recommendation for drop control is to make the initial application before the first sound fruit start to drop. NAA at 10 mg L^{-1} is frequently used and drop control typically lasts for 7-9 days. A second supplemental application can be made to extend the period of drop control up to 14 days. NAA may advance ripening on apples when used to control drop. The amount of stimulation depends on the cultivar used, the weather following application, time interval between application and harvest and the amount NAA applied. Higher rates of NAA, very warm weather (mid-80°F (30°C)) and a delay in the harvest of treated fruit are likely to advance ripening and lead to reduced storage life. Drop control following NAA application can be seen within 2–3 days of application unless significant drop was occurring at the time of application.

NAA is sometimes used in conjunction with ReTain®. Since ReTain® requires 10–14 days for drop control to be initiated, NAA may be included in the application for near-term drop control. A minimum of a half rate of ReTain® is required to counteract the tendency for NAA to advance ripening. Recently, Robinson et al. (2010) and Yuan and Carbaugh (2007) reported addition of NAA in the second part of a ReTain® split application that was reported to enhance

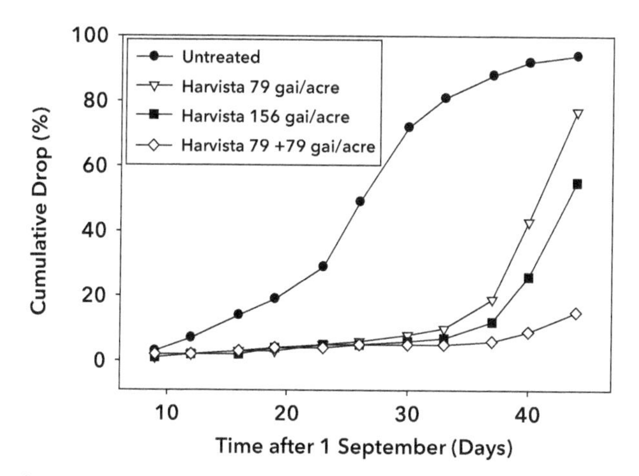

Figure 11 Influence of single and double application of Harvista™ on pre-harvest drop of 'McIntosh' apples. The higher rate of Harvista™ controls drop more effectively than a lower rate. However, a second application of the lower rate of Harvista™ made 2 weeks after the first resulted in drop control better than the higher rate applied only once.

© Burleigh Dodds Science Publishing Limited, 2019. All rights reserved.

the drop control of a ReTain®-alone application. An alternative approach to drop control by NAA preloading was reported by Marini et al. (1993). Application of several low rates of NAA (5 mg L^{-1}) starting 4 weeks before anticipated harvest resulted in excellent drop control of 'Delicious' apples that was equal to or better than a higher rate made at the traditional time nearer harvest. This approach has been useful on 'Delicious' but it has not been successful on more ethylene-sensitive cultivars such as 'McIntosh'.

8.2 Advance fruit ripening and increase red colour

As normal harvest approaches and seasons start to change, there is usually heightened interest by consumers to purchase first-of-the-season apples. However, this interest usually precedes the time that the fruit are ripe, palatable and ready to harvest. There is the temptation by orchardists to harvest fruit early when it lacks adequate colour, is starchy, acidic and tannic in order to satisfy this demand. However, selling unripe and inferior quality fruit would depress the market and discourage customers from making return purchases. Fortunately, there are PBR combinations available to allow the harvest of good quality fruit that have the varietal characteristics of fruit harvested later during the normal harvest season.

8.2.1 Drop control when using ethephon

Whenever ethephon is applied to apple trees a drop control PBR must be used as well. Ethephon causes ethylene evolution when absorbed by the tree and increased endogenous ethylene production is stimulated which advances ripening resulting in increased pre-harvest drop, especially on drop-prone cultivars. There are three drop control compounds available but only ReTain® and NAA have been extensively evaluated to be used with ethephon. In situations where ethephon is being applied primarily to advance ripening, NAA is the drop control most often selected since the advanced ripening caused by NAA is not necessarily detrimental and the cost is less. One application of NAA at 10 mg L^{-1} will generally control drop for about 7 days and a second application may extend drop control to 12–14 days. Some orchardists choose to delay application of NAA until 4 days after ethephon application thus extending the period of drop control without having to make a second application. This strategy comes with some peril since poor weather may prevent a grower from making the application in a timely manner and drop starts before the NAA can be applied or become effective.

In situations where ethephon is applied primarily to increase red colour but storage of harvested fruit is planned then the use of ReTain® may be the appropriate choice to control drop. Since ReTain® delays ripening as well as

© Burleigh Dodds Science Publishing Limited, 2019. All rights reserved.

red colour development, a reduced rate may be appropriate in this situation. It is suggested that reduced rate be used (½ pouch, 166 gal $acre^{-1}$) and it should be applied 2–2.5 weeks before ethephon is applied.

8.2.2 Ethephon to advance ripening and increase red colour

Ethephon is the primary PBR used to advance fruit ripening. It should be applied between 1 and 3 weeks before normal harvest at rates between 62.5 and 300 mg L^{-1}. Increased red colour will be noted 5–7 days after application. Flesh softening may accompany red colour development. The closer the application is made to the time of normal harvest, the shorter the time required to observe a red colour and maturity response. If warm cloudy weather follows application, advanced ripening may occur with little or no increase in red colour development.

Ethephon-treated fruit should be monitored carefully following harvest since a reduction in storage life should be expected. Factors that may shorten the storage potential of fruit include use of high rates of ethephon, a long interval between the time of application and harvest, high temperatures while fruit is on the tree, the drop control compound selected and a delay in the time to cool the fruit down so that the internal temperature is reduced to 0°C. Therefore, treating apples with ethephon will mean early harvest of fruit, which will result in reduced yield due to smaller fruit size. A rule-of-thumb is that a fruit increases in size about 1% per day provided it remains on the tree.

8.2.3 Increase red colour with ethephon

There are some geographical locations where fruit do not develop good red colour and there are years when warm and cloudy weather persists resulting in poorly coloured fruit. In situations such as this, some growers wish to enhance red colour without appreciable advancing of ripening. In situations such as this a rate of ethephon of between ¼ and ½ pint of ethephon per 100 gal (75–150 mg L^{-1}) should be applied 7–10 days prior to the anticipated time to harvest. Care should be taken to monitor fruit maturity to make sure that fruit do not become too ripe on the tree. In recent years more emphasis has been placed on using starch testing as guide to help determine maturity and when to harvest fruit.

9 Improving fruit appearance and shape

Appearance of fruit often plays a dominant role in purchasing decisions made by customers and greatly impacts the price that growers receive from the packing house. Good cultural techniques and management of the tree are

© Burleigh Dodds Science Publishing Limited, 2019. All rights reserved.

essential prerequisites but frequently PBRs may be required to improve fruit appearance even when management is excellent.

9.1 Fruit russet control

Russet is a physiological disorder that occurs during the early stages of fruit development. Rapidly dividing epidermal cells increase in size faster than the cuticle can expand. This leads to the cuticle cracking, exposing and damaging some of the epidermal cells below. As a result, cambium that is formed produces corky periderm russetted areas (Faust and Shear, 1972). Although not the only cause of russetting, water plays a critical and central role in its formation, as it tends to be more pronounced under conditions of high humidity or frequent rain or dew. These conditions lead to an apple developing a thin cuticle that is prone to cracking under high turgor conditions (Faust and Shear, 1972). Some varieties are also genetically pre-disposed to develop russet.

Foliar sprays of GA_{4+7} can either eliminate fruit russetting or reduce its severity (Taylor, 1975). When applied near petal fall at rates of 25–250 mg L^{-1} GAs are known to inhibit bud formation in flowers even when used at moderate rates (Tromp, 1982). By making multiple sprays (generally 4) of GA_{4+7} at a rate of 10 mg L^{-1}, fruit russetting can be reduced while having minimal effect on return bloom (Miller, 1988). High rates may improve russet control but the increased control may not be justified because of its adverse effect on return bloom (Eccher and Boffelli, 1981).

Scarf skin is a disorder of apples that is characterized as fruit having a grey flecking and/or a milky appearance on the surface. Its presence on the fruit leads to a reduction in grade and therefore lower returns to the grower. Treatment of susceptible cultivars with multiple applications of GA_{4+7} (as described for russet) will reduce the incidence of scarf skin but it may not reduce the incidence to commercially acceptable levels (Ferree et al., 1984).

9.2 Fruit shape

Fruit shape can be a significant component when customers choose an apple to purchase. Cultivars that are blocky and have prominent calyx lobes are perceived by customers as being preferred. Apple elongation and calyx lobe development are promoted by low temperatures during the very early fruit growth stage (Shaw, 1914). Cool nights during early fruit development characterize the growing conditions in Washington. The Washington apple industry has been able to market shape, especially for 'Delicious' as a quality parameter. Cool nights cause increased production of endogenous GAs. Abbott Laboratories developed a proprietary product, Promalin®, to elongate the shape of 'Delicious'. It contains equal quantities of GA_{4+7} and 6-benzyladenine

© Burleigh Dodds Science Publishing Limited, 2019. All rights reserved.

(BA) and this was registered in 1979. The application of this product near bloom allows growers in other regions that do not have cool temperatures at night to produce elongated 'Delicious' with prominent calyx lobes. Unrath (1974) demonstrated that when Promalin® was applied at 25 mg L^{-1} at bloom time, fruit length/diameter ratio and weight increased. Using GA rates above 50 mg L^{-1} could result in a reduction in return bloom. GA_4 and GA_7 increase fruit elongation but GA_7 is known to reduce flower bud formation more than GA_4 (Tromp, 1982).

10 References

Abbaspoor, M., Teicher, H. P. and Streibig, J. C. 2006. Effect of root-absorbed inhibitors on Kautsky curve parameters in sugar beet. *Weeds Res.* 46:226-35.

Bangerth, F. 1978. The effect of a substituted amino acid on ethylene biosynthesis, respiration, ripening and preharvest drop of apple fruit. *J. Am. Soc. Hort. Sci.* 103:401-4.

Bangerth, F. 2004. Internal regulation of fruit growth and abscission. *Acta Hort.* 636:235-48.

Barry, C. S. and Giovannoni, J. J. 2007. Ethylene and fruit ripening. *J. Plant Growth Reg.* 26:143-59.

Batjer, L. P. and Hoffman, M. B. 1951. Fruit thinning with chemical sprays. U.S. Dept. Agr. Circ. 867. US Department of Agriculture, Washington, DC.

Beyer Jr., E., Morgan, P. W. and Yang, S. F. 1984. Ethylene, pp. 111-26. In: M. B. Wilkins (Ed.), *Advanced Plant Physiology*. Pitman Publishing Ltd., London, UK.

Boller, T., Herner, R. C. and Kende, H. 1979. Assay for the enzymatic formation of an ethylene precursor 1-aminocylopropene-1-carboxylic acid. *Planta* 145:293-303.

Bound, S. A., Close, D. C., Jones, J. E. and Whiting, M. D. 2014. Improving fruit set of 'Kordia' and 'Regina' sweet cherry with AVG. *Acta. Hort.* 1042:285-92.

Bukovac, M. J. 1963. Induction of parthenocarpic growth of apple fruit with gibberellins A_3 and A_4. *Bot. Gaz.* 124:191-5.

Bukovac, M. J. 1965. Some factors affecting the absorption of 3-chlorophenoxy-α-propionic acid by leaves of peach. *Proc. Am. Soc. Hort. Sci.* 87:131-8.

Bukovac, M. J. 1973. Foliar penetration of plant growth substances with special reference to tree fruit. *Acta Hort.* 34:69-78.

Bukovac, M. J. 1980. The performance of growth regulators when applied in low volume sprays. *Proc. Michigan State Hort. Soc.* 93:63-7.

Bukovac, M. J. 1985. Spray application technology: A limiting factor in growth regulator performance. *Proc. Wash. State Hort. Soc.* 81:170-4.

Byers, R. E. 1993. Controlling of bearing apple trees with ethephon. *HortScience* 28:1103-5.

Byers, R. E. 2003. Flower and fruit thinning and vegetative: fruiting balance, pp. 409-36. In: D. D. Ferree and I. J. Warrington (Eds), *Apples: Botany, Production and Uses*. CABI Publishing, Cambridge, MA.

Byers, R. E., Lyons, C. G. and Hill, C. H. 1971. Base gallonage per acre. *Virginia Fruit* 60:19-23.

Byers, R. E. Barden, J. A. and Carbaugh, D. H. 1990a. Thinning of spur 'Delicious' apples by shade, terbacil, carbaryl, and ethephon. *J. Am. Soc. Hort. Sci.* 115:9-13.

© Burleigh Dodds Science Publishing Limited, 2019. All rights reserved.

Byers, R. E., Barden, J. A., Young, R. E. and Carbaugh, D. H. 1990b. Apple thinning by photosynthetic inhibition. *J. Am. Soc. Hort. Sci.* 115:14–19.

Chan, D. and Cain, J. 1967. The effect of seed formation on subsequent flowering in apple. *Proc. Am. Soc. Hort. Sci.* 91:63–8.

Cleland, R. E. 1969. The gibberellins, pp. 49–81. In: Wilkins, M. B. (Ed.), *Physiology of Plant Growth and Development*. McGraw-Hill, Maidenhead, UK.

Clouse, S. D. 1996. Molecular genetic studies confirm the role of brassinosteroids in plant growth and development. *Plant J.* 10:1–8.

Clouse, S. D. and Sasse, J. M. 1998. Brassinosteroids: Essential regulators of plant growth and development. *Annu. Rev Plant Physiol. Plant Mol. Biol.* 49:427–51.

Creelman, R. A. and Mullet, J. E. 1997. Biosynthesis and action of jasmonates in plants. *Annu. Rev. Plant Physiol. Plant Mol. Biol.* 48:355–81.

Crocker, W., Hitchcock, A. E. and Zimmerman, P. W. 1935. Similarities in the effect of ethylene and the plant auxins. *Contrib. Boyce Thompson Inst.* 7:231–48.

Davies, P. J. 1995. *Plant Hormones. Physiology, Biochemistry and Molecular Biology*. Kluwer Academic Publishers, Dordrecht, The Netherlands, 833pp.

Dennis, F. G. 2002. Mechanisms of action of apple thinning chemicals. *HortScience* 37:471–4.

Eccher, T. and Boffelli, G. 1981. Effects of application of GA_{4+7} on russeting, fruit set, and shape of 'Golden Delicious' apples. *Sci. Hort.* 14:307–14.

Evans, J. R., Evans, R. R., Rogusci, C. L. and Rademacher, W. 1999. Mode of action, metabolism and uptake of BAS 125W prohexadione-calcium. *HortScience* 34:1200–1.

Fan, X., Mattheis, J. P. and Fellman, J. K. 1998. A role for jasmonates in climacteric fruit ripening. *Planta* 204:444–9.

Fan, X., Blakenship, S. M. and Mattheis, J. P. 1999. 1-methylcyclopropene inhibits apple ripening. *J. Am. Soc. Hort. Sci.* 124:690–5.

Faust, M. and Sherer, C. B. 1972. Russeting of apples, in interpretive review. *HortScience* 7:233–5.

Ferree, D. C., Ellis, M. A. and Bishop, B. L. 1984. Scarf skin on 'Rome Beauty': time of origin and influence of fungicides and GA_{4+7}. *J. Am. Soc. Hort. Sci.* 109:422–7.

Forshey, C. G. 1982. Branching responses of young apple trees to applications of 6-benzylamino purine and gibberellins A_{4+7}. *J. Am. Soc. Hort. Sci.* 107:128–32.

Fukuda, H. 1997. Tracheary element differentiation. *Plant Cell* 9:1147–6.

Greene, D. W. 1999. Tree growth management and fruit quality of apple trees treated with prohexadione-calcium. *HortScience* 34:1209–12.

Greene, D. W. 2002. Chemicals, timing, and environmental factors involved in thinner efficacy on apple. *HortScience* 37:9–13.

Greene, D. W. 2006. An update of preharvest drop control of apples with aminoethoxyvinylglycine (ReTain). *Acta Hort.* 727:311–19.

Greene, D. W. and Autio, W. R. 1994. Notching techniques increase branching of young apple trees. *J. Am. Soc. Hort. Sci.* 119:678–82.

Greene, D. W. and Bukovac, M. J. 1971. Factors influencing the penetration of naphthaleneacetamide into leaves of pear. *J. Am. Soc. Hort. Sci.* 96:240–6.

Greene, D. W. and Costa, G. 2013. Fruit thinning in pome and stone-fruit: State of the art. *Acta. Hort.* 998:93:101.

Greene, D. W. and Lord, W. J. 1978. Evaluation of scoring, limb spreading, and growth regulators for increasing flower bud formation and fruit set on young 'Delicious' apple trees. *J. Am. Soc. Hort Sci.* 103:208–10.

© Burleigh Dodds Science Publishing Limited, 2019. All rights reserved.

Greene, D. W. and Schupp, J. R. 2004. Effect of aminoethoxyvinylglycine (AVG) on preharvest drop, fruit quality and maturation of 'McIntosh' apples. II. Effect of timing and concentration relationships and spray volume. *HortScience* 39:1036-41.

Greene, D. W., Krupa, J. and Clements, J. 2000. Effects of simulate rain following ReTain application on preharvest drop and fruit quality of McIntosh apples. *Fruit Notes* 65:57-60.

Greene, D. W., Lakso, A. N., Robinson, T. L. and Schwallier, P. 2013. Development of the fruit growth model to predict thinner responses to apple. *HortScience* 48:584-7.

Greene, D. W., Krupa, J. and Autio, W. 2014. Factors influencing preharvest drop of apples. *Acta Hort.* 1042:231-5.

Greene, D. W., Crovetti, A. J. and Pienaar. J. 2016. Development of 6-benzyladenine as an apple thinner. *HortScience* 51:1448-51.

Jacobs, W. 1979. *Plant Hormones and Plant Development*. Cambridge University Press, New York, NY.

Jones, K. M., Koen, T. B., Oakford, M. J. and Bound, S. A. 1990. Thinning 'Red Fuji' apples using ethephon at two timings. *J. Hort. Sci.* 65:381-4.

Katsumi, M. 1991. Physiological modes of brassinolide action in cucumber hypocotyl growth, pp. 246–54. In: H. G. Cutler, J. Yokota and G. Adams (Eds), *Brassinosteroids Chemistry, Bioactivity and Applications*. American Chemical Society, Washington, DC.

Lakso, A. N., Robinson, T. L. and Greene, D. W. 2006. Integration of environment, physiology, and fruit abscission via carbon balance modeling - Implications for understanding growth regulator response. *Acta Hort.* 727:321-6.

Lakso, A. N., Robinson, T. L. and Greene, D. W. 2007. Using an apple tree carbohydrate model to understand thinning responses to weather and chemical thinners. *N. Fruit Quart.* 15(3):17-20.

Letham, D. 1963. Zeatin, a factor inducing cell division from *Zea Mays*. *Life Sci.* 2:569-73.

Marini, R. P. 1996. Chemically thinning spur 'Delicious' apples with carbaryl, NAA, and ethephon at various stages of fruit development. *HortTechnology* 6:241-6.

Marini, R. P., Byers, R. E. and Sowers, D. L. 1993. Repeat applications of NAA control preharvest drop of 'Delicious' apples. *J. Hort. Sci.* 68:247-53.

McArtney, S. J. 1994. Exogenous gibberellin affects biennial bearing and the fruit shape of 'Braeburn' apple. *New Zeal. J. Crop Hort. Sci.* 22:343-6.

McArtney, S. J. and Obermiller, J. D. 2012. Comparison of the effects of metamitron on chlorophyll fluorescence and fruit set in apple and peach. *HortScience* 47:509-14.

McArtney, S. J. and Obermiller, J. D. 2015. Effect of notching 6-benzyladenine and 6-benzyladenine plus gibberellin $A_4 + A_7$ on budbreak and shoot development from paradormant buds on the leader of young apple trees. *HortTechnology* 25:233-7.

McArtney, S. J., Palmer, J., Davies, S. and Seymour, S. 2006. Effects of lime sulfur and fish oil on pollen tube growth, leaf photosynthesis and fruit set in apple. *HortScience* 41:357-60.

McArtney, S. J., Greene, D., Robinson, T. and Wargo, J. 2014. Evaluation of GA_{4+7} plus 6-benzyladenine as a frost-rescue treatment for apple. *HortTechnology* 24:171-6.

Milborrow, B. V. 1984. Inhibitors, pp. 76-110. In: M. B. Wilkins (Ed.), *Advanced Plant Physiology*. Pitman Publishing Ltd., London.

Miller, S. S. 1988. Plant bioregulators in apple and pear culture. *Hort. Rev.* 10:309-401.

Miller, C., Skoog, F., Okumura, F., Von Saltza, M. and Strong, F. 1955. Kinetin, a cell division factor from deoxyribonucleic acid. *J. Am. Chem. Soc.* 77:1392.

© Burleigh Dodds Science Publishing Limited, 2019. All rights reserved.

Moore, T. C. 1989. *Biochemistry and Physiology of Plant Hormones*. 2nd Edition. Springer-Verlag, New York, NY, 300pp.

Musacchi, S. and Greene, D. 2017. Innovations in apple tree cultivation to manage crop load and ripening. In: K. Evans (Ed.), *Achieving Sustainable Cultivation of Apples*. Burleigh Dodds Science Publishing, Cambridge, UK.

Peck, G. M., Combs, L. D., DeLong, C. and Yoder, K. 2016. Precision apple flower thinning using organically approved chemicals. *Acta Hort*. (*In press*).

Phinney, B. O. 1983. The history of gibberellins, pp. 19–52. In: A. Crozier (Ed.), *The Biochemistry and Physiology of Gibberellins Vol. 1*. Proeger Publishing, New York, NY.

Price, C. E. 1982. A review of factors influencing the penetration of pesticides through plant leaves, pp. 237–52. In: D. F. Cutler and C. E. Price (Eds), *The Plant Cuticle*. Linnaeus Soc. Symp. Ser. No. 10, Academic Press, London, UK.

Reid, J. B. and Howell, S. H. 1995. Hormone mutants and plant development, pp. 448–85. In: P. J. Davies (Ed.), *Plant Hormones: Physiology, Biochemistry and Molecular Biology*, 2nd ed. Kluwer Academic Publishing, Boston, MA.

Robinson, T., Hoying, S. and Reginato, G. 2008. The tall spindle planting system: principles and performance. *Acta Hort*. 903:571–80.

Robinson, T. L., Hoying, S., Ungerman, K. and Kviklys, D. 2010. ReTain combined with NAA controls preharvest drop of 'McIntosh' apples better than either chemical alone. *N. Y. Fruit Quart*. 18(3):9–13.

Robinson, T., Lakso, A., Greene, D. and Hoying, S. 2013. Precision crop load management. *N. Y. Fruit Quart*. 21(2):3-9.

Robinson, T. L., Hoying, S., Miranda Sazo, M. and Rufato, A. 2014. Precision crop load management Part 2. *N. Y. Fruit Quart*. 22(1):9–13.

Robinson, T. L., Lakso, A. N., Greene, D., Reginato, C. and De R. Rufato, A. 2016. Managing fruit abscission in apple. *Acta Hort*. 1119:1–14.

Rudell, D. R., Fellman, J. K. and Mattheis, J. P. 2005. Preharvest application of methyl jasmonate to 'Fuji' apples enhances red coloration and affects fruit size, splitting, and bitter pit incidence. *HortScience* 40:1760–2.

Savaldi-Goldstein, S. 2006. Brassinosteroids, pp. 617–34. In: L. Taiz and E. Zeiger (Eds), *Plant Physiology*. Sinauer Associates Inc., Sunderland, MA.

Schwallier, P. G. 1996. *Apple Thinning Guide*. Great Lakes Publishing Co., Sparta, MI.

Schwallier, P. G. and Irish-Brown, A. 2015. Predicting apple fruit set model. *N. Y. Fruit Quart*. 23(1):15–18.

Schupp, J. R. and Greene, D. W. 2004. Effect of aminoethoxyvinylglycine (AVG) on preharvest drop, fruit quality and maturation of 'McIntosh' apples. I. Concentration and timing of dilute application of AVG. *HortScience* 39:1030–5.

Shaw, J. K. 1914. A study of variations in apples. *Mass. Agr. Expt. Sta. Bull*. 149:29–36.

Stover, E. W. and Greene, D. W. 2005. Environmental effects on the performance of foliar applied plant growth regulators: A review focusing on tree fruit. *HortTechnology* 15:214–21.

Taylor, B. K. 1975. Reduction of apple skin russeting by gibberellin A_{4+7}. *J. Hort. Sci*. 50:169–72.

Tromp, J. 1982. Flower-bud formation in apple as affected by various gibberellins. *J. Hort. Sci*. 57:277–82.

Unrath, C. R. 1974. The commercial implication of gibberellin A_4 A_7 plus benzyladenine for improving shape and yield of 'Delicious' apples. *J. Am. Soc. Hort. Sci*. 99:381–4.

© Burleigh Dodds Science Publishing Limited, 2019. All rights reserved.

Watkins, C. B. 2003. Principles and practices of postharvest handling and stress, pp. 585-615. In: D. C. Ferree and I. J. Warrington (Eds), *Apples, Botany, Production and Uses*. CABI Publishing, Cambridge, MA.
Wertheim, S. J. 2000. Developments in chemical thinning of apple and pear. *Plant Growth Reg.* 31:85-100.
Westwood, M. N. and Batjer, L. P. 1960. Effects of environment and chemical additives on absorption of naphthaleneacetic acid by apples leaves. *Proc. Am. Soc. Hort. Sci.* 76:16-29.
Williams, M. W. 1980. Retention of fruit firmness and increase in vegetative growth and fruit set of apples with aminoethoxyvinylglycine. *HortScience* 15:76-7.
Williams, M. W. and Billingsley, H. D. 1970. Increasing the number and crotch angles of primary branches of apple trees with cytokinins and gibberellic acid. *J. Am. Soc. Hort. Sci.* 95:649-51.
Williams, M. W. and Edgerton, L. J. 1981. Fruit thinning of apples and pears with chemicals. U.S. Dept. Agr. Bul. 289. US Department of Agriculture, Washington, DC.
Williams, M. W. and Stahly, E. A. 1969. Effect of cytokinins and gibberellins on shape of 'Delicious' apple fruit. *J. Am. Soc. Hort. Sci.* 94:17-19.
Wismer, P. T., Proctor, J. T. A. and Elfving, D. C. 1995. Benzyladenine affects cell division and cell size during apple fruit thinning. *J. Am. Soc. Hort. Sci.* 120:802-7.
Yang, S. F. 1969. Ethylene evolution from 2-chloroethylphosphonic acid. *Plant Physiol.* 44:1203-4.
Yarushnykov, V. V. and Blanke, M. W. 2005. Allocation of frost damage to pear flowers by application of gibberellin. *Plant Growth Reg.* 45:21-7.
Yoder, K. S., Peck, G. M., Combs, L. D. and Byers, R. E. 2013. Using the pollen tube growth model to improve apple bloom thinning for organic production. *Acta Hort.* 1001:207-14.
Yuan, R. and Carbaugh, D. H. 2007. Effect of NAA, AVG, and 1-MCP on ethylene biosynthesis, preharvest drop, fruit maturity, and quality of 'Golden Supreme' and 'Golden Delicious' apples. *HortScience* 42:101-5.

© Burleigh Dodds Science Publishing Limited, 2019. All rights reserved.

Printed in the USA
CPSIA information can be obtained
at www.ICGtesting.com
JSHW011551010724
65693JS00006B/121